WHAT BECAME OF THE CROW?

THE INSIDE STORY OF THE GREATEST GOLD DISCOVERY IN HISTORY

ROBERT MORIARTY

Also by Robert Moriarty
    Basic Investing in Resource Stocks
    Nobody Knows Anything
    The Art of Peace
    Crap Shoot
    Exposed!
    Entrapped!

What Became of the Crow? © 2021 by Robert Moriarty. All rights reserved. No part of this book may be used or reproduced in any manner whatsoever without written permission, except in the case of brief quotations embodied in critical articles or reviews.

First Edition

Editing by Jeremy Irwin, jc9cz@yahoo.com

Library of Congress Cataloging-in-Publication Data has been applied for.

ISBN 978-1-716-17094-2

DEDICATION

To my beloved wife Barbara, who can never know how much I miss her every day. Life is so boring without her.

A mine is a hole in the ground. The discoverer of it is a natural liar. The hole in the ground and the liar combine and issue shares and trap fools. — The *Detroit Free Press*, 1881

It seems to me that it's up to all of us to try to tell the truth, to say what we know, to say what we don't know, and recognize that we're dealing with people that are perfectly willing to, to lie to the world to attempt to further their case and to the extent people lie of, ultimately they are caught lying and they lose their credibility and one would think it wouldn't take very long for that to happen dealing with people like this. — Donald Rumsfeld, U.S. Secretary of Defense, 2004

The difficult we do immediately; the impossible takes a little longer. — Charles Alexandre de Calonne (1734–1802), Finance Minister to King Louis XVI

As geology is essentially a historical science, the working method of the geologist resembles that of the historian. This makes the personality of the geologist of essential importance in the way he analyzes the past. — Reinout Willem van Bemmelen (1904–83), Dutch geologist

CONTENTS

	Introduction	1
1	Blowing up Osama Bin Laden in a boot	5
2	There is no rush like a gold rush	11
3	What became of the crow?	13
4	In the beginning . . .	19
5	Mark Creasy and the Pilbara	25
6	It takes two to tango	31
7	Where I fit in	43
8	Placer Gold on the South Island	49
9	Swilling $1,200-a-bottle wine	61
10	Galliard and the early years	83
11	2014 brings production opportunity	107
12	When academics loot and plunder	115
13	Coming up with a new way to mine	119
14	Quinton discovers a rich gold mine	131
15	Sniffs of gold at Karratha	139
16	Quinton flubs the first assays at Purdy's	171
17	There is crazy and there is Pilbara crazy	181
18	The peak, and then the long decline	195
19	Novo set the bar too high at Karratha	203
20	Kirkland Lake attempts a coup	209
21	Taking advice from keyboard commandos	221
22	Moving forward towards production	227
23	Millennium goes teats up	255
24	It's not rocket science	263
	References	265

INTRODUCTION

WITH AN UTTER LACK OF ACCURACY, mining has been described as the art and science of extracting minerals from the ground at a profit. Alas, it's not an art. It's hardly a science. Profit rarely shows its face.

When young students attend school to learn about geology and the mining industry, no one tells them about the idiotic bureaucrats seemingly determined to stop each project at every possible turn. The instructors fail to inform the fledglings about partners who lie, cheat, and steal, or employees who seek at every opportunity to sabotage their company's operations in the hopes of taking it over themselves. Minor issues, such as natives angered at being cheated by every mining venture that ever came along, are not raised. While the locals can't make a company succeed, they can make it fail.

Geology and mining is a minefield where every step you take can blow your leg off. The path forward features booby traps and trip wires, and on each side there are hidden pits filled with sharp punji stakes waiting to pierce the unwary. Competitors will steal your claim markers when they are not stirring up the locals to obstruct everything you do.

Everyone fabricates; everyone deceives and seems dedicated to making sure you fail. Your own technical staff will present their own theories, all of which will disappoint. A geologist without a pet theory that can never succeed is a rare specimen.

When markets go up, everyone wants to throw money at ventures. In an instant those same markets may tumble, and promises of funding turn to dust as demand for the product evaporates.

If geology students had even a lick of common sense they

would pursue an honest profession, such as the law or prostitution, where at least someone finds satisfaction.

We are all liars. We tell lies to maintain harmony in our lives. Without the mercy of lies we would be constantly at war with our governments, and even with our significant other. Lying can be a good thing.

Your bride comes home after shopping for a new pair of jeans and asks politely, "Do these pants make my butt look big?"

She's really asking two questions. One has to do with the fit of the jeans; the other is about how much you care for her. There is a correct answer and there is a right answer.

"Darling, the last time I saw an ass that big was on a southbound hippo at the San Diego zoo in 1975." That's an example of a correct answer.

"Darling, the only way your bottom could look better would be if you were wearing nothing at all. Those pants look wonderful on you. Have you lost weight?"

That's the right answer. You may have told a little white lie about the fit of the jeans but you have reassured your wife that you love her.

We want to be lied to. Can you imagine a TV preacher or politician actually telling you the truth or a newsletter writer? They always lie; if they didn't, no one will contribute to their new aircraft fund or vote for them or buy their pulp. We don't want to know the truth. Most newsletter writers specialize in telling people what they want to hear. Their subscribers will never make any money, but the newsletter writers will.

Over the past twenty years I have probably visited about five hundred mining projects. I've seen just about everything. Gold, silver, lead, coal, copper, diamonds, a few oil wells, rare earths, uranium, iron ore. Name a metal or a mineral resource or a country, and I have probably visited a project site and seen it.

Sometimes I've visited the same project twice, with different companies owning it.

I'd guess I get lied to about 75 percent of the time. Many times those briefing me will tell a whopper that even a dunce couldn't miss. The smart ones manage not to mention what you really need to know, if you are to understand whether or not they are sitting on a deal-killer. These are falsehoods by omission. What they fail to mention is what you really need to know.

I like being lied to 75 percent of the time. Twenty-five years ago my wife Barbara and I ran a computer business where we were lied to 100 percent of the time. It's actually quite refreshing for someone to tell me the truth now and again. It's so different.

I read a lot. I'll read anything that comes to hand; I don't much care what the subject is. But naturally, since I am associated with mining, I read as many books as I can about resources, investing, and mining companies. Occasionally I even write a book about those same things.

I've never read a book about mining that told the truth, the whole truth, and nothing but the truth. They all lie, either by omission or commission. I'm old enough now not to really give a shit I can pay my bills, so the purpose of this book is to tell the whole story about mining, leaving nothing out. Well, maybe just a tiny little bit.

This is the mostly true story of the greatest gold discovery in world history. I was a fly on the wall almost from the beginning. The driving force behind it happens to be my best friend.

I'm not sure I'm his best friend, but he's certainly my best friend. I love his wife and his grandkids as well, and doesn't that sort of make us family?

Once upon a time. . .

CHAPTER 1
BLOWING UP OSAMA BIN LADEN IN A BOOT

THE MASSIVE BRE-X GOLD FRAUD [1] peaked in 1997. The price of gold had been fairly steady for years, but gold shares were in the process of completing a long advance. It needed only a pinprick to pop the bubble. Bre-X proved to be more of a whaling harpoon. The fraud ended up affecting the lives of thousands of men and women in the mining industry.

It all began in 1993, when a tiny Canadian company named Bre-X Minerals bought a mining project in Borneo called Busang; it was located on the Busang River. Bre-X did the typical ground survey and basic exploration in advance of drilling, and then started a major drill program in 1995.

At first they claimed to have a gold deposit of two million ounces, which was nice, but not enough to grab the attention of a major gold company. But by the end of 1995 they claimed to have discovered thirty million ounces, making Busang one of the biggest gold projects in history.

They continued to drill, and each hole apparently hit incredible intervals of gold at high grades. By 1996 Bre-X was telling the world that the deposit was really sixty million ounces, and when 1997 rolled around it was up to an incredible seventy million ounces of bonanza grade gold, with every drill hole showing gold.

The sharks moved in for the kill. Placer Dome attempted a takeover but failed. The government of Indonesia announced that the project was far too large for a tiny Canadian junior. Fortunately, a solution was at hand. With the help of President Suharto's daughter, Bre-X could share the project with Barrick Gold. But then it was decided that Freeport-McMoRan Copper &

Gold would run the mine and Bre-X would maintain a 45 percent interest, negotiated by Suharto's son Sigit Hardjojudanto, who would receive a billion-dollar cut for his participation.

It probably should have occurred to someone that in all of recorded mining history, no project had ever gone from two million ounces of gold to seventy million ounces in three years. It also would have been worth having someone look at the gold.

Bre-X's story was that the reported lengths and grades were the result of analyzing diamond drill core extracted from hard rock. But even a cursory examination of the gold particles, with a simple hand lens of the type possessed by every geologist, would have revealed clear signs of it being nothing more than placer gold, mined by local Indonesians and sold to someone at Bre-X.

In short, the company salted the assays. Everyone involved, from the management of Bre-X to Barrick to President Suharto to Placer Dome, forgot to employ their common sense due to being blinded by the idea of instant riches. The reader will find that shortcoming repeated throughout this book; the lure of gold blinds people to reality.

Canadian bureaucrats then came up with an entirely new set of regulations called National Instrument 43-101, designed to set standards on how and when assays and resources would be announced. It was an attempt to put a crimp on the fraud so common in the gold mining business. But wherever there is gold, someone will come up with a new variation on some old tried and proven scam.

Everyone in mining would like to pretend that Bre-X was some sort of outlier that had never happened before and couldn't happen again. But that's hardly true. There are 1,200 to 1,500 junior resource companies in Vancouver and Toronto. In time, ten or twenty of them might come up with a nice economic deposit. The rest are peddling some variety of moose pasture. Management's primary objective most often is to continue to collect paychecks, so they can live the lifestyle they have come to crave.

There are a hundred variations of Bre-X going on at any given time. Bre-X just got carried away, and dragged many others down with them.

Within a year of the fraud being exposed, virtually all gold exploration work worldwide stopped. Dr. Quinton Hennigh, PhD (one of the main characters in this story) was working for Newcrest Mining at the time. The company told its exploration staff that they would be laid off, but they were given severance pay. Quinton left mining in late 1998 and started a new career.

In October of 2001, a young mathematics and science teacher who was new to the job took his sixth-grade class on a school trip, to see an operating gold mine in the foothills of western Colorado. The bus took them to Nederland, where the young sprouts got to visit an authentic gold mine.

It had been agreed that the owner and operator, Tom Hendricks, would give a talk about some subject or other, and then ask the kids questions about what he had just tried to teach them. Those who answered successfully would be given a treat. It might be a sample gold specimen or an old used drill bit or an antique can that had once held blasting caps.

Tom raised a hand and told the kids that whoever could answer his next question would get to light the dynamite. Being twelve years old, the male sixth-graders all thought that would be just wonderful. A few of the young ladies thought it sounded pretty spiffy as well, this being Colorado.

Tom asked the question. A lad named Gavin answered it correctly. So at the end of the tour, Tom said that it was time for

Gavin to light the dynamite.

Tom took an old rubber boot and drew a picture of Osama bin Laden on the side. He went to the explosives locker and picked up a quarter stick of dynamite. Carefully, he crimped an eighteen-inch length of pyrotechnic fuse into a blasting cap. Then he punctured the end of the dynamite stick and inserted the blasting cap. He dropped the now fused and ready explosive into the boot and looked around for his assistant, who was to have the honor of lighting the fuse.

While twelve-year-old boys think lighting dynamite sticks is a great adventure, the first time they try it, they tend to get a bit nervous. Tom reached into his pocket to pull out a Zippo lighter before handing it to Gavin, whose hand was visibly shaking. The young teacher had to steady him so he could start the fuse burning.

Tom held the boot carefully, and as the fuse burned down he continued to lecture the children. It was October, and the weather changed. Snow started to fall. The school was in Longmont, down in the flats east of the Rockies. The kids dressed for fall, not winter. Some of them were now shaking from the cold and wanted nothing more than to get back on the bus and return to school.

Eventually the group dispersed and ran for a warm spot on the bus. When the burning fuse was at the proper length, Hendricks tossed the boot and its explosive over the top of the building behind him.

Twelve-year-old kids have short attention spans. Lighting the fuse was interesting, as was learning how to blow things up. But nothing happened after he threw the boot over the building.

Until it blew up with a massive explosion.

Since 9/11 had been the center of their world for the previous month, the kids were especially sensitive to explosions.

All of them immediately fell to the ground. About half of them proceeded to pee in their pants.

The teacher looked at Hendricks and said, "Tom, those are twelve-year-olds. The dynamite trick might be a little much for them. I think we should wait until I bring out the eighth-grade students. I'm sure they would love the idea of blowing up bin Laden in a boot."

The story of Gavin helping Tom Hendricks to blow up Osama bin Laden in a boot is a tale still often repeated today in Longmont, Colorado, a small town where not much happens. That teacher is still remembered fondly as an inspiration. With the slight exception of the school administration.

Quinton Hennigh was of course that young school teacher. After leaving Newcrest in 1998 he had taken a job at the Twin Peaks Charter School in Longmont. He found himself mostly teaching the children of cattle ranchers. Math and science were hardly a priority, but 4-H [2] was. When Quinton was hired to teach both subjects to grades 6–8, the students were scoring at the 50 percent level in mathematics and a dismal 26 percent in science. Within one year those same students, with their new teacher, rose to 76 percent statewide.

Years later, several of his students went on to attend and graduate from the Colorado School of Mines. One student that Quinton first met and taught when he was in the seventh grade, Jacob Nuechterlein, completed his PhD at age 27. He went on to found one of the world's leading companies designing high-technology materials for 3D printing in the way of advanced metals, composites and ceramics. Quinton and I are both investors in the company.

Quinton likes to tell of the time he blew up the school science lab. As part of his teaching he would get some liquid nitrogen, put a couple of tea spoons of the stuff in a plastic 7-Up bottle and put the cap on loosely. One day he tightened the cap a bit too tightly. Normally the nitrogen turned to gas and made an interesting sight, acting like a gas-powered rocket around the classroom. Alas, on this occasion it blew up, taking part of the classroom roof with it.

I much prefer the story of how he and Tom Hendricks blew up the boot. Anyone can blow up a 7-Up bottle, but it takes a real man to blow up Osama bin Laden. I'm certain the kids who participated still remember that particular adventure.

But while it was interesting for Quinton, his heart was in gold discovery, not the detonation of footwear.

CHAPTER 2
THERE IS NO RUSH LIKE A GOLD RUSH

HOW CAN THERE BE A GIANT GOLD DISCOVERY without an exciting story of just how the gold was found?

What would the California Gold Rush of the nineteenth century be without the exciting tale of James Marshall peering down into the waters of the American River as he built a water-powered sawmill for John Sutter? He spied small flakes of what looked to be gold.

It was gold.

And the rush was on, drawing tens of thousands of would-be prospectors and miners, all believing that the streets of California were surely paved with gold. All you had to do in order to get rich was to pick it up. The California Gold Rush literally settled the west, as fortune hunters from around the world converged on California and the supposed opportunity for rapid wealth.

Bret Harte and Mark Twain left their stories of the California Gold Rush. By reading their tales millions of people still participate in the adventure and heartbreak that accompany every gold rush.

At the dawn of the twentieth century, the world underwent another gold rush as thousands made the difficult and dangerous journey to the wilds of the frozen Yukon in their search for wealth. Many spent as long as eighteen months in their journey. Most, over half, would take one look at Dawson and turn right back around. The trip proved so difficult that the journey itself became the destination.

Jack London wrote a number of exciting books and stories about the people involved, and even their dogs. Robert Service told tales that still entertain today, mixing a few facts, a lot of fiction and a large dose of humor in his descriptions of the Klondike, and what miners went through in their search for gold.

Any form of instant riches will always attract a crowd but gold has a special niche of its own. The first time you see a flake of gold in the bottom of a gold pan you are often hooked for life. And reading about it can be just as mesmerizing.

CHAPTER 3
WHAT BECAME OF THE CROW?

JAMES WITHNELL STARTED THE FIRST Pilbara gold rush by accident in early 1888. All he was trying to do was to protect his lunch from a pesky and persistent crow with a hankering for home cooking.

Mallina Station, on the Sherlock River, has sparse vegetation at the best of times. The boiling summer days of January 1888 had James thinking that the Western Pilbara basin in Western Australia would make a great summer retreat for the Devil, should he ever seek a more temperate clime than that of Hell. The Devil would find the Pilbara in midsummer to be moderate. Mad dogs and Englishmen believe it's hotter than Hell and are not far wrong, with average January temperatures in the Basin as high as 39 degrees Celsius or 102 Fahrenheit.

James wiped sweat off his brow as he and his brother Harding chopped wood for their cooking fire. With one eye he watched a crow hopping about in a dance around his dinner pail; it seemed to be planning to filch his lunch. The crow recognized James and appreciated boiled beef sandwiches.

Raising cattle and sheep in the scrubland of Western Australia, or WA, was proving to be hard work, with little return. It was hot and dusty work. A decent lunch in the field made the task worth doing. But there would be no lunch if James were to turn his back on the crow at the wrong time.

The crow hopped closer to the lunch pail and looked around, as if on a summer vacation on the beach at a crow festival. James saw an iron-stained cobble at his feet and reached down for it, to throw it at the bird. He picked it up, and in an instant forgot all about both the crow and the lunch.

He weighed the rock in his hand; tossed it into the air and caught it. It seemed particularly heavy; far too heavy to be quartz or ironstone. Jimmy fell to his knees and picked up and weighed other rocks, to see if they too were unusually dense. He searched within a ten-foot circle and found half a dozen iron-stained pieces of quartz that were heavier than such rocks usually are.

Could it be gold?

There were rumors of gold having been found in the Pilbara Basin a few years before. And the newspapers were filled with stories of the riches being found in the south of Africa, in the British colony there. If gold nuggets littered the ground about Mallina, he might get rich.

Meanwhile the crow planned on munching Jimmy's lunch as it thought, "Humans are strange creatures, lacking keen appreciation for what is valuable in life. You can find rocks everywhere, but a fine lunch . . . "

James emptied his lunch pail onto the ground and received the crow's gratitude. He piled his newfound stones into the pail, and the two brothers ran home. They quickly hitched their oxen to their wagon and set off for Roebourne,[3] seventy-three miles west on the dusty track, to show their find to their father.

The pair made good time through the heat and were in the tiny town a day later. Roebourne was the nearest town of distinction to the Mallina Station. It possessed scant distinction, but did have the nearest telegraph terminal.

Their parents, John and Emma Withnell, ran a store there and provided cartage services to the other settlers. Their house served as a hub to the two hundred residents of the town, and when necessary provided a venue for church services, until such time as a church could be built.

"Da, Da, I think I've found gold," James exclaimed as he

emptied his pail onto their table.

John Withnell shook his head at the naivety of his son. "Don't be daft," he began. "That's quartz with iron staining." He lifted one of the stones to prove his point. It did seem unusually dense, and there were stringers of something running through the quartz crystal.

"Son, you may just be right. It is substantial. Too stout to be just a simple pebble. We should show this to Lieutenant-Colonel Angelo, and see what he says."

Lt-Col. Edward Angelo [4] had fought with valor in the Crimean War, earning multiple decorations. Later, after retiring from running the Western Defense Force in 1886, he served as the Government Resident at Roebourne. He was said to be a bit stuffy, but would understand what a gold discovery might mean to the Pilbara Basin.

A large gold nugget

The excited trio of Withnell men rushed into his tiny office in the town square. "Colonel Angelo," John began, "I think James and Harding have found some gold. These stones are too heavy to be just quartz. Can you tell us what you think?" James poured his parcel of rocks onto the colonel's otherwise orderly desk.

As the Withnell family had helped settle Roebourne since their arrival in 1863, the colonel always had time for them. The matriarch, Emma Withnell, [5] was the first white female settler in northwest WA. She came from the Hancock family, a later generation of which would start the Pilbara iron mining industry. [6]

Eventually, as a result of the ensuing gold rush, Roebourne would become the largest city between Darwin and Perth, some nine hundred miles to the south, down the coast.

"Tell me the whole story, James. Where and how did you find these rocks? What makes you think they might be gold? They certainly don't look like much," Angelo said as he twirled the mustache that made him resemble a walrus.

"Well, sir, we were cutting wood for the fire about a hundred yards from the cabin. I put my lunch bucket down so I could work. As I chopped, a crow made a move to steal me tucker. I picked up a rock to chuck at it. It seemed very heavy. I looked around and found more rocks that looked about the same. They are all a lot stouter than I'm used to. I thought they might be gold so I came here with them to show Da."

"Hmmm," Colonel Angelo said in the pompous way so common to former military officers, "we must send a telegraph to Governor Broome in Perth and tell them we think you have found gold. O'Malley! Come in here. I want a telegraph sent to Perth at once."

His assistant, O'Malley, came into the office and took a seat, ready to transcribe a telegram to the government seat at Perth.

"Now, O'Malley, it is extremely important to get this off to Perth straight away, so bloody well pay attention," the Resident began. O'Malley had a tendency to wool gather and could be a burden at times.

"Begin. Gold discovered at Mallina. Jim Withnell picked up a stone to throw at a crow." Angelo noticed O'Malley examining his pencil, and stopped speaking.

"Sorry sir, but my pencil needs a topper. Give me a minute." He added, "I have that so far."

He rushed off, believing he had the essence of Colonel Angelo's message. He returned in a few minutes with a sharpened pencil, ready to continue. "Sir, I apologize for the delay, but I have sent that telegraph off to Perth."

Colonel Angelo's face turned crimson as he realized O'Malley believed he had been given the entire message intended for Perth. "You dunderhead," Angelo stammered, "there was more to come." Under his breath he muttered, "Bloody Irish."

At the time the Irish in the colony of Australia were only a few generations old, since arriving on the island as convicts. They disliked the English almost as much as the English despised them. Under his breath O'Malley mumbled, "Bloody pommies." [7]

The Governor of Western Australia, Sir Frederick Broome, in Perth, found himself baffled by the contents of the truncated telegraph.

He quickly responded, "What became of the crow?"

CHAPTER 4
IN THE BEGINNING...

WHEN THE GOVERNMENT IN PERTH finally understood that Lt-Col. Angelo was reporting a possible new find of gold, a new Australian gold rush began. Within six weeks, fifty miners had staked claims on the Mallina Station. In some areas the surface portion of the rich quartz vein was twenty feet thick.

Under British law, the mineral rights to a deposit belonged to the Crown. Even if James Withnell had the rights to the station he lived on, he had no more right to any minerals there than anyone else. Within weeks prospectors had chased the gold-rich quartz veins over the entire area, staking claims within the limits rigidly enforced under British law. Every man had an equal opportunity to take a chance on becoming instantly wealthy. Thousands took advantage of the laws governing gold mining.

In 1863, when Emma Withnell and family had settled in Roebourne, the total population of Western Australia was a tiny 12,300. (That's 12,300 people in an area about one-quarter of the size of the U.S.)

In the next twenty-five years, up to the time of the gold discovery in 1888, the population of WA doubled to 25,800. In the next three years it almost doubled again, to 49,790, as men and a few women were drawn to the idea of instant wealth. Naturally, the male/female ratio was skewed. In WA, males outnumbered females by two to one. However, then as now, there were a few female gold diggers around.

More gold rushes in WA followed, at Cue in 1891 and Coolgardie in 1892. The fabulous gold wealth from Kalgoorlie in 1893 drew tens of thousands of adventurers from around the world to Australia, and WA's population had increased to

184,100 by 1901. In just ten years it had almost quadrupled.

The Pilbara discovery was reported in local and national news journals. West Pilbara district drew miners by the thousands, each eager to seek fame and fortune. By August of 1888 the prospectors had searched as far east as the Oakover River, two hundred miles east of Roebourne. They found not only more outcrops of quartz laden with gold, but also numerous gold nuggets in unconsolidated alluvial gravels near water courses.

The gold in the Pilbara region of WA has many similarities to that in the world's largest gold area, the Witwatersrand District of South Africa. You will find hard rock gold in quartz veins such as at Mallina, just as you will find veins of gold in the Wits.

Much of the gold in the Pilbara is located in gravel reefs, some of them consolidated into hard rock, some more loosely packed. Even then, over a hundred and thirty years ago, the miners realized that much of the alluvial gold had its origin in the stacked conglomerate sequences, or reefs, one or two meters thick. In the Wits there are areas with as many as fourteen reefs of varying richness and thickness.

Finally, in both regions there was alluvial gold washed out of the reefs and vein systems and concentrated by both erosion and water.

Clearly, the Pilbara Basin was rich in gold. From the alluvial gravels around Nullagine at the eastern end of the Basin, all the way to the shores of the Indian Ocean, one hundred and forty miles to the northwest, the gravel was rich in coarse nuggets.

But there were problems. Water was always a problem, there being either too little of it or too much. In the monsoon season the entire area flooded. For most of the rest of the year there was a lack of water for processing. With the technology of the day, working alluvial gravels required lots of water to process masses of material. The availability of water proved to be either a feast or a famine.

The technology of 1888 pretty much dictated the type and level of mining. At first, with rich seams of gold at surface and nuggets in every stream and riverbed, a man could become rich within days. There are newspaper reports from December of 1888 about prospectors who found nuggets as large as fifty-eight ounces while using a dry blower to separate the gold from the dross.

The *Western Mail* of March 22, 1890 carried an item about Nullagine: "James Carey has worked out his claim, getting between 300 and 400 ounces of gold. Beaton gold 150 ounces from one load of wash dirt obtained at a depth of 27 feet."

More reports from Nullagine in April of 1890 suggest that several nice patches of gold nuggets had been discovered, with some men taking as much as "20 ounces in a few days, but the majority are only getting a few penny-weights per day." (A pennyweight, or dwt, is about 1.5 grams. There are 20 dwt to a troy ounce.)

In November of 1890 it was reported that at the Turner River, "gold was found at this spot some months ago and it is estimated about three hundred ounces have since been unearthed." Further, "The first few men who arrived hit some very rich patches ... and were getting about 5 ounces per diem each, but that was soon worked out."

While gold rushes are certainly exciting to read about, with tales of instant or near-instant fortunes being made, technology and vision always determine their potential in the end. Northern WA had little in the way of infrastructure and no permanent rivers suitable for navigation. All goods had to be brought in by sailing vessels.

The first Pilbara gold rush made a good story but didn't

produce all that much gold, compared to other areas of the colony of Australia. Official records of Pilbara gold production begin in 1889 and show only 11,170 ounces of gold being produced that year. That increased over the next few years to a maximum of 20,526 ounces in 1899, after which there was a rapid decline.

The primary importance of the Pilbara gold rush was that it acted as a catalyst for prospectors and miners to seek other richer and more amenable areas to mine in the territory. What the tens of thousands of miners wandering the length and breadth of the Pilbara Basin didn't realize was that all the gold there probably had the same origin. It took a PhD candidate from the Colorado School of Mines to realize that the Pilbara Basin almost certainly has a similar endowment of gold to that of the Witwatersrand.

The glory years of the Pilbara and West Pilbara districts were between 1888 and 1900. Production declined continuously after that, although small-scale mining would continue for another hundred and twenty years.

Coins to be used as currency were a big problem in the colony for many years. All of the gold mined in WA would be taken to the offices of the Royal Mint in Perth and exchanged for British sovereigns and half-sovereigns minted in the Sydney and the Melbourne Royal Mints. The Sydney Mint began producing sovereigns in 1855. The Melbourne Mint followed in 1872. It took until 1899 for there to be sufficient gold produced in WA to justify the cost of building a third mint. The Perth Mint opened in June of that year.

The combined production from the Pilbara and West Pilbara districts, from the time of James Withnell's discovery up until 1901, totaled about 190,000 ounces of gold. Compared to the 760,805 ounces from the Murchison district, in the Mid West region of WA, the Pilbara didn't look like much. Mount Magnet, also in the Mid West, yielded 474,100 ounces over the same period. East Coolgardie and the Coolgardie district had produced a whopping 4.5 million ounces of gold by 1901.

Records indicate that only 163 men were working in gold mining in the West Pilbara and Pilbara districts in 1900. In comparison, East Coolgardie counted 5,903 laboring in the gold mines there.

At the time gold was money and provided the liquidity for the economy to function. It's vital to understand just how important gold was to the economy of Australia, and how great the contribution made by Western Australia was, from 1888. The Perth Mint's records show that more than 106 million gold sovereigns were struck between 1899 and 1931, when Britain abandoned the gold standard. That is more than twenty-six million ounces of gold, or nearly $50 billion worth in today's U.S. dollars and at today's gold price. That wealth built the entire infrastructure of the territory and eventually led to the mining of iron becoming a mainstay of the economy of WA.

I'm going to drop in a hint here. The massive iron deposits of WA contributed $4.4 billion in Aussie dollars in royalties to the government in 2018. WA holds an incredible 29 percent of the entire world's iron reserves. It provided 39 percent of the world's iron supply in 2018. The gold came from the same place as the iron, but no one understood that until recently.

The gold was there in the Pilbara, but most of it stayed there for over a century more until a prospector's dream was combined with a geologist's vision and with advances in technology. That formidable combination began to turn a hot and barren landscape covered with spinifex grass and gangly shrubs, with a few brave kangaroos showing themselves at dusk, into — well, let's call it a gold mine.

CHAPTER 5
MARK CREASY AND THE PILBARA

PRETTY MUCH EVERYTHING IN LIFE, up to and including sex, is easier and more fun with assistance from another person. When you do things by yourself and just for yourself, it's often called masturbation.

The discovery and advancement of the biggest gold project in world history was largely the result of putting two particular people together. Many others took part, including me, but it required the brilliance of (a) the man whom Australians consider to be their greatest prospector, and (b) a somewhat chubby PhD geologist with an interesting theory that he had been gnawing on for years, like a starving dog with a bone.

Seeing them alone, or even together, wouldn't impress anyone. One is tall and skinny, the other not. You could dress both of them for $30 at the nearest Goodwill used clothes shop and make them look sharper and get change back. However, if you talk with either of them for five minutes, about anything, you will soon realize you are in the company of genius. Neither is aloof but neither is keen on fools.

Britain's so-called victory in World War II bankrupted the country. Subsequent mismanagement and poor decision-making by the Labour government in power from July 1945 until its defeat by the Conservatives in 1950 ensured the continuation of the food rationing imposed during the war. Bread rationing began in 1946, after the war had ended! It was 1954 before all rationing ended. Even cheese production took another thirty years to recover to pre-war standards.

The food-rich islands of Australia and New Zealand donated a form of CARE aid packages to Britain in the years after the war. For a young man born in 1944 and growing up in the gray Britain of the 1950s, the Pacific colonies must have looked like a combination of an adventure playground and a food paradise.

Mark Creasy, the young man in question, devoured everything Jack London wrote about the Klondike gold rush. He filled his head with ideas of someday finding a gold mine and becoming instantly wealthy. After graduating in 1964 from the Royal School of Mines, part of Imperial College in London, the twenty-year-old set off for Australia to turn his dream into reality.

At the time, Australia realized it needed immigration, especially of young people with skills and education. The government would pay for their transportation and set them up with housing and a job when necessary.

Mark's first job was in a Queensland coal mine, as a mining engineer. There may be some area of geology and mining more boring than coal, but I have yet to find it.

His interests soon turned to opal mining. For a time he worked at Broken Hill before going into real-life prospecting for opals in New South Wales and Queensland. While opals are exciting for a young mining guy, there are no great fortunes to be made in their discovery. Mark's thoughts kept returning to the tales of Jack London and the riches of the Klondike.

That encouraged him to move in 1972 to Perth, capital of Western Australia, to start a highly successful career as a prospector of gold and then nickel. In an article published by the Australian Prospectors and Miners Hall of Fame, Mark is quoted as saying about Jack London, "It didn't take long to realize that London knew next to nothing about the game."

Mark became one of those overnight successes that are twenty years in the making. "All I know is, I starved for years. Looking for minerals is a fascinating occupation and it gets into you, if you like. It's intellectually stimulating."

His luck began to change in 1976 when, using a metal detector, he found a 46-ounce gold nugget near Mount Magnet. He discovered another monster 86-ounce specimen near Laverton in 1977. The pair of nuggets, along with some of others of smaller size, sold for a $60,000 windfall that would finance his prospecting for the next few years.

Mark believed the Yandal greenstone belt would be prospective for gold. In 1978 he began true prospecting, taking stream samples and doing geochemical surveys to narrow down a potential area. It took many years of hard work and near-starvation wages before he found success in 1991.

Mark tells a tale of how one day he drove his beat-up Land Cruiser from his work in the field into the nearest town for supplies. As he sat in a bar having a sandwich and drinking a beer, he glanced at a newspaper and realized it was his birthday. He had forgotten.

In interviews over the last few years Mark laments how government bureaucracy has changed the nature of mining in WA, and not for the better. When he began fifty years ago, a miner's permit cost fifty cents. When a prospector pegged some potential ground, the exploration rights would be granted in six weeks or less. Those same permits take years to be granted today. That delay tends to kill junior mining companies. Independent prospectors are long gone.

Prospectors prior to the early 1970s largely ignored the Northern Yandal greenstone belt, even though it's but twenty-five miles from the Wiluna gold camp. In 1983 Mark filed claims on what would eventually become the Jundee gold mine. He formed a company with a few others, but they failed to raise the funds necessary to advance the project. They lost the tenements.

Creasy reclaimed the ground in 1985 and went to Chevron

with his idea of forming a joint venture to advance the Jundee–Nimary ground. Chevron took a year to prepare an agreement, and then at the last possible moment decided to pull out of Australia entirely. Mark was left holding the bag, with the required work not completed and a statutory requirement to give up 50 percent of the licensed ground.

Mark had found gold nuggets on the Jundee ground but none on the Nimary property, so he dropped the Nimary portion. The Department of Mines stipulated a ninety-day waiting period before a project could be restaked. A bare week before the ninety days expired, a company called Hunter Resources pegged the Nimary ground from right underneath Mark's feet. Hunter promptly found four gold deposits and built a mine.

During 1989 and 1990 Creasy completed a geophysical survey on the remaining Jundee ground and followed up with eighty-three short rotary air blast (RAB) holes.[8] Only four of them showed a grade higher than one gram of gold per ton of rock. In 1991 he talked Great Central Mines into a joint venture. By 1995 the JV had drilled hundreds of RAB holes and outlined a resource of 1.3 million ounces of gold.

In 1995 Mark Creasy caught the brass ring for the first time. Or should I say the gold ring? Great Central Mines purchased the rest of the Jundee tenements from him, for $117 million. (Mining land properties in Australia are referred to as tenements. It's a block of land, nothing more.)

That payoff rang the bell for everyone from miners to ordinary citizens to the taxman. That most definitely set the fox amongst the chickens. You see, at that time the government of WA understood it needed prospectors to seek and find new deposits. So when a prospector sold a claim, the proceeds were entirely tax-free. Mark collected his check for $117 million and deposited it in his bank account. It was all his to spend.

The Australian public threw a hissy fit. How outrageous that someone might profit by the sweat of his labor and be allowed to keep the money he had earned, instead of handing a big chunk of it to the taxman, so the government bureaucrats could spend it wisely — far more wisely than the person who had earned it. The law was changed, and prospectors could no longer make money without Big Brother taking a giant bite out of it.

The public hated Mark because he had made money and not shared it with them. His fellow prospectors despised him as well, blaming him for causing the change in the law. Sometimes you just can't win. All he had to console himself with was a check with a lot of zeros on it.

Australians will never admit it, but they think Big Brother is just great. They love socialism and hate anyone who makes money through honest work.

There was one little clause in the agreement between Creasy and Great Central that everyone except Mark passed over at the time. It would become important many years later. Mark demanded and got what is called an ROFR: a Right Of First Refusal. If at any time in the future the Jundee mine were to be put up for sale, Mark had the right to match any offer made.

As an asset of Great Central Mines, Jundee later went through a series of owners. But it was the company changing hands, which from a legal point of view isn't the same as the mine changing hands. The ROFR would come into play only if the mine should ever be sold by itself.

Mark now had a lot of money to explore with. I've visited his office in Perth several times. Call it the ultimate man cave. Mark buys all kinds of things related to mining, just because they interest him. He has the greatest collection of gold nuggets in Australia, as well as magnificent specimens of all sorts.

He even has what is the Gutenberg Bible of mining: the copy of *De re metallica* ("On the Nature of Metals"),[9] a sixteenth-century book, that had once belonged to a mining engineer working in the gold mines of Western Australia, named Herbert Hoover.

Hoover and his wife translated that particular book from Latin into English for the first time in 1912 and left many handwritten comments in it. Hoover is remembered in Australia as a mining engineer; in the U.S. he is better known for being its thirty-first president.

Mark's collection of mining memorabilia is easily the biggest and best in Australia, perhaps even in the world. Along with his hundreds of rare and valuable books on mining, someday a museum is going to get a good deal.

In the back of his mind he has always believed that the jewelry box, the big Kahuna of gold as it were, was in the Pilbara Basin. No one had ever made any big discoveries there but Mark believed something giant was hidden there.

Technology and basic metallurgical issues were big problems. Even the first discovery at Mallina Station by James Withnell came loaded with difficulties. A five-stamp mill was brought in to crush the quartz, but the gold was associated with so much antimony that it couldn't be amalgamated with mercury.

Where there was free gold in the form of nuggets — at Egina, Nullagine, and Pilbara Creek — there was either too much water or too little. Mark suspected the Basin was loaded with gold, but it would take someone with a different variation of his vision to figure out how to retrieve it.

CHAPTER 6
IT TAKES TWO TO TANGO

MARK CREASY NEEDED THE RIGHT PARTNER to move the Pilbara forward to where he believed it could go. Da Vinci required someone to block his canvases prior to painting, and to mix paints for him. He might be the master, but even masters need someone to cook their meals. Everything worth doing requires the efforts of multiple people.

Mark continued to prospect, and kept tossing money into prospective juniors. But his crowning glory, of putting together the greatest gold discovery in history, wasn't going to happen until he could enlist the aid of someone with similar vision and intelligence.

That partner wandered into his life in 2005 in the form of a 39-year-old junior geologist, albeit a PhD, working for Newmont Mining. That would be Quinton Hennigh, the other heroic figure in our saga.

When Hollywood gets around to making the movie from this book, the head of casting will encounter a difficulty. The only person in the world even faintly qualified to play the part of Quinton Hennigh, alas, is Quinton Hennigh. Like Mark Creasy, he's unique. Intelligent, visionary, determined, and willing to listen to most, even the idiots. He's a far better man than me. I can't stand idiots and the world is filled with them.

Quinton began his mining career at the age of four, when his grandfather took him into the Colorado foothills, starting west of Denver to search old mining dumps for interesting specimens and maybe even gold. He commuted between his birthplace in Missouri, where his parents lived, and where his father was a university professor, and Longmont, Colorado, where he stayed half the year with his grandparents.

His boyhood was filled with the idea of someday making a giant gold discovery. Maybe even the biggest in world history.

Imagine that.

At the age of eighteen he got his start working on a real gold mine, the Caribou Mine [10] near Nederland, Colorado. Tom Hendricks, the owner of the mine, hired him to scrub the toilets. You might say Quinton got his start in mining at the very bottom. He spent the summers of his high school and college years learning the ropes as he carefully cleaned toilets.

Quinton completed his bachelor's degree, then his master's, and then went to the Colorado School of Mines, the *crème de la crème* of U.S. mining schools, to study for his PhD.

Now, when you undertake a PhD course you should pick an interesting subject for your thesis. Quinton wanted to determine the true origin of the gold in the world's largest gold field.

The largest at that time, that is.

That was the Witwatersrand deposit in South Africa. Goldfields funded his work and he made good progress, but in 1993, due to financial constraints, Goldfields pulled its financing.

One of Hennigh's professors, Richard Hutchinson, found merit in the 27-year-old student and pressured him into working on Hutchinson's theory that black smokers or geothermal vents were the source of gold and pyrite in the sedimentary Wits Basin. Black smokers are underwater hot water pipes spewing out mineral-rich super-hot fluids. The metal-rich fluids resemble smoke coming out of a chimney.

In a moment of serendipity Quinton dissolved samples of the Wits conglomerate in hydrofluoric acid. When he did so, he kept finding particles of carbon. He then came across a paper written in 1978, on the subject of how carbon would trap particles of gold. [11] Not quite. The carbon and carbonaceous material didn't trap the gold particles; it caused the gold to precipitate out of solution. It was Quinton's Eureka moment.

One of the things they teach in geology courses, or which you pick up for yourself on your seventh or eighth site visit, is that gold likes carbon, and will attach itself to carbon under the right conditions. Seawater at the time was reduced, to use a technical term. That means there was no oxygen in it. It was loaded with hydrogen sulfide. Hydrogen sulfide literally eats gold in a reduced environment.

Believing that the gold-bearing solution somehow caused gold to attach itself to carbon, Quinton set out to work backwards. He knew the age of the Wits; it's about 2.6 to 2.9 billion years old. He did some research to determine the chemical composition of water at that time.

Single-cell creatures in the seawater began to produce oxygen and the chemistry and pH of the water changed over time. Since carbon and gold get along so well together, the gold attached itself to the carbon.

I'll backtrack for a moment and give readers the basic chemistry course I was given, I think between site visits 101 and 118.

Hot water deep in the Earth, with various chemical compositions, will dissolve anything. The high-pressure superheated water is always seeking lower pressure, in the same way that winds go from high-pressure systems to low-pressure systems. The fluids generally seek to rise. They find their way to the Earth's surface through fissures in the rock. When the temperature or the chemistry or the pressure of the water changes, minerals will precipitate out.

All minerals are found everywhere. But to have an economic deposit you need a concentration of some mineral.

I mentioned earlier that all geologists have a pet theory. It's like when you were a kid and had a pet hamster. You have your hamster (or your theory) and you come to love it because it's all yours. But you not only love your hamster; you must also pee on everyone else's hamster, or theory. This is for reasons of NIH, or Not Invented Here.

If you came up with a theory, and told everyone about it, and they all loved it, you would inevitably discover that it was a bullshit theory. For a theory to be valid, many people must pee on it.

Quinton found that most people didn't greatly care for his theory, primarily because they hadn't thought of it first. But if you have a good theory, time tends to prove it was right, eventually. You know its time has come when the guys who spent so much time peeing on your hamster now want to steal it in the dead of night, and then claim it was their hamster in the first place.

Those in academia are famous for pinching the ideas they made so much fun of for so long. Naturally there is some of that in our story.

Quinton ended up doing his PhD thesis on ore deposits in Portugal. How interesting. Be still my beating heart.

But he never lost sight of his goal. That was to make the biggest gold discovery in history, and to find it in a basin where the gold had precipitated out of solution. He would also need the right dance partner.

Geologists are a lot like hookers. Eventually they work every corner in town. Quinton learned all that you would ever want to know about scrubbing toilets by working with Tom Hendricks off and on for five years. Then he worked for Homestake for a couple of years, and for the United States Geological Survey (USGS) for just long enough to learn that people take government jobs only because they are too lazy to work and too nervous to steal. He had a short stint with AUR Resources before working for the best of the best, Newcrest, from 1994 to 1998.

The Bre-X scandal in 1997 caught the gold market at a high. As the price tumbled for years, geologists were laid off by the score, Quinton among them. For five years he taught middle school mathematics.

He survived the years spent facing classrooms of teenagers with hormones raging before he signed on at Newmont in 2004. During his period in exile the industry had changed totally. In the 1980s and 1990s the majors did their own exploration work. Gradually, by the early years of this century, the big companies had laid off most of their exploration staff.

Newmont still believed it should be doing basic exploration work. After hearing Quinton's theory about how gold had been precipitated out of salt water 2.8 billion years ago, Newmont hired him. They told him to form a team and go find some paleoplacer deposits. His group spent eighteen months putting together data from all over the planet. They located ninety-two basins of the right age and with indications of gold. The biggest and best was the Pilbara Basin in Western Australia.

It was said of George Washington that he was "First in war, first in peace and first in the hearts of his countrymen."[12] But George married Martha, a widow. Even George Washington wasn't first in everything.

Neither was Quinton with his precipitation theory.

He may have been third or fourth, but writers as early as 1896 were offering the same theory. In a privately published book entitled *Witwatersrand Gold – 100 Years*, printed in a limited edition by the Geological Society of South Africa in 1986, we read this on page 31 (I've put the good bit in bold text).

De Launey in his book 'Les mines d'or du Transvaal' of

1896 also considered a fluviatile origin for the conglomerates. Pretorius (Depositional Environment of Witwatersrand Goldfields) published the following summary of the essential elements of the hypothesis: de Launey wrote that it was possible that the Witwatersrand strata could represent widespread fluviatile sediments deposited on torrential deltas distributed over an alluvial plain such as that which he had observed in Lombardy in Italy however, his preference was for a marine beach origin because of the vast extent of the formations, the frequent occurrence of shingle type pebbles and because the Witwatersrand conglomerates occupied an "intermediate stratigraphic position between two formations known to contain marine fossils — the Bokkeveld beds and the carboniferous limestones.

De Launey agreed with his compatriot Garnier that the gold was secondary and had been precipitated from the ocean waters into the unconsolidated conglomerates.

And on page 32:

Prof August Prister read his paper, 'Notes on the Origin and Formation of the Witwatersrand Auriferous Deposits' to the Geological Society in May 1898. He approved of the generally accepted sedimentary origin of the conglomerates but proposed an entirely new mode of mineralization in these beds. His hypothesis demanded that the Witwatersrand sedimentary epoch be completed in toto, after which the entire supergroup subsided and was flooded by oceanic waters holding gold and pyrite in solution. These waters infiltrated the sediments before final consolidation to precipitate the gold and pyrite in the pore space of the conglomerates. Because the Main Reef group of conglomerates had a well-nigh impervious layer of mud or silt in its footwall, the seepage of this solution ceased at that depth and there precipitated most of its dissolved gold and pyrite. For that reason the Main Reef group of conglomerates were more heavily mineralized with gold and pyrite. The overlying pebble beds received less as the solution was actively passing through on the way down to the Main Reef horizon and the dilutionary effect of rain water reduced precipitation at the shallower depths. Prister stated that the entire process would be repeated as crustal movements brought about renewed uplift and subsidence.

Newmont's exploration manager in Australia, Brian Levet, somehow discovered Quinton at the company office in Denver. They began to talk about the Pilbara and the opportunity there. Brian pointed out that a prospector in Perth held a commanding position in land claims in the Pilbara. His name was Mark Creasy. Quinton suggested to Newmont that he go to Perth and talk to Mark.

There were a few snags.

Actually there were a lot of snags.

By this time Quinton had formalized his theory, passed it around Newmont, and found a lot of support. He just needed the right partner and land package to move forward.

Quinton flew to Perth with Brian. They met with Mark. On behalf of Newmont, Brian signed a confidentiality agreement with Mark that basically forbade Newmont from talking to anyone else about doing deals. It was a dumb move on Brian's part to agree to that because it tied Newmont's hands without getting anything in return. (Quinton knew nothing about the terms of the agreement.)

Mark agreed with Quinton about the potential source of the gold in the Pilbara. He would not only sell all of his Pilbara gold claims to Newmont, but would add a giant land position in South Africa he happened to hold, in another one of Quinton's ninety-two basins of the right age. All he wanted in return was a check for a squillion dollars.

But it would be a cold day in hell before Newmont would write a check for a squillion dollars for kangaroo pasture, no matter where it was located or how well decked-out in gold the kangaroos were.

I mentioned in the beginning of this book that your own people will do dumb things to queer your deal. Brian Levet had signed a confidentiality agreement with Mark, promising that Newmont wouldn't deal with anyone else. Then he went to Wedgetail Exploration (the predecessor of Millennium Minerals, of whom more later) and signed another confidentiality agreement with them in order to discuss doing a deal on the Beatons Creek property.

The Australian mining sector is comprised of a bunch of people who gossip more than a clutch of old ladies with blue hair at a church social. Five minutes after Brian signed the agreement with Wedgetail, Mark knew about it. It annoyed him, to the extent that he was no longer about to do a deal with Newmont.

That first meeting between Mark Creasy and Quinton was in October of 2005, and they got along famously. Quinton traveled up to the Pilbara to take samples, and came back with thirty samples averaging six grams of gold per ton. His fieldwork verified or at least supported his theory. But Mark had first demanded a check that Newmont wasn't about to write, and then didn't want to deal with Newmont at all because in his view, Brian Levet was double-dealing.

Actually, the only dumb thing Brian did was to sign that restrictive agreement with Mark. That was stupid. In mining you have to deal with whoever will do a deal on reasonable terms. Mark was being unreasonable and didn't want Newmont dealing with anyone else.

Meanwhile, Quinton knew he had found the Promised Land. Unfortunately his dance partner wanted nothing to do with his employer. Thereafter, Brian kept fobbing Quinton off with one excuse after another about why the project wasn't moving forward, even though he knew Mark wasn't about to cut an agreement with Newmont. That went on for a couple of years until Quinton accidentally stepped on his dick.

It was the very best thing that he could have done.

Quinton made another trip to WA in April of 2006 to pick up another three hundred samples. This time they averaged four grams of gold per ton. That's economic, and it convinced Quinton of two things. One was that the Pilbara Basin was loaded with gold, and the other was that it was his destiny (that he had been working on since he was four years old) to put it all together.

In May or June of 2006 Quinton happened to run into Pierre Lassonde, the president of Newmont, in the company parking lot in Denver. He gave Pierre the elevator pitch on the project. Pierre asked him who controlled it, and Quinton told him. Pierre is not a big fan of Mark Creasy, and he told Quinton that if he ever ran into difficulty, Pierre would get involved and make the deal happen.

Brian kept stalling Quinton. Eventually Quinton e-mailed

Pierre, asking for his help.

There are two important things to learn here.

One is that people typically will refuse to admit they have screwed up. All business is a series of trial and error. You make a lot of decisions. Some work, some don't. Stick with those that work and chuck the rest. Running a company is nothing more than solving problems one at a time. But solving problems requires first knowing about them, and there's the rub. Because people won't tell you they screwed up. I've run half a dozen projects and said the same thing to every single person who ever worked for me.

I say this to them. "Everyone makes mistakes. It's no big deal. But to solve a problem, I have to know about it. Please, please don't hide shit from me. I hate surprises. If you don't solve a problem promptly it will only get worse. Tell me about every problem you run into and I'll try to fix it."

Of course, everyone I said that to totally ignored me. I could repeat it until the cows come home. It wouldn't make any difference. In the mining business, the people who work for you just love to surprise you with problems they found out about six months ago, and which have grown to giant size in the interim.

Running any business, but I think especially a mining business, means constantly being whacked on the side of the head with a giant problem that was a simple issue months ago when it first appeared, but was covered up and ignored because people just won't admit to fucking something up. They would far rather give you a horrible surprise at the worst possible time.

The second lesson to be learned is that when the big boss gives you his e-mail address with his direct line and tells you to get in touch with him personally if you can't sort something out, he doesn't really mean it. Throw away his business card or you will find yourself in the shit. The boss wants the chain of command firmly adhered to, even if that chain is comprised of blithering idiots who try to hide all their fuck-ups.

Luckily for the discovery of gold in Australia, and for the mining business in general, Quinton needed to prepare to pay for three kids soon to enter college but was dead broke. He had married the lovely Heather in 1989 and they soon started producing infants once they got the hang of it. Quinton had just entered his fifth decade. He had little in the way of savings and frankly wasn't being paid much at Newmont.

Quinton contacting the boss directly annoyed everyone at Newmont. It just wasn't done. Meanwhile, every former taxi driver or drill crew supervisor in Vancouver had now started a junior mining company or two and was raking in the money. You didn't need good sense or any special skill set. You just needed a sufficiently good patter to convince a resource mutual fund to back you and throw some money your way.

I'm not exaggerating when I say that money was being thrown at some of the biggest bullshit mining projects in history. There were areas of Nevada and Newfoundland where there were so many drill holes that it looked like Swiss cheese. You needed to strap a 2 × 4 to your ass in case you tripped and fell in.

Like Daniel before him, Quinton Hennigh saw the writing on the wall and made his exit from Newmont just before the lions got to him. It was a brilliant move on his part to leave in March of 2007. The junior market would surely appreciate his talents far more than the majors did.

But not for a minute did he lose sight of his goal. He knew who he wanted to sign his dance card. Mark Creasy was equally determined to work with someone, but he was free of the financial pressure that Quinton faced. Quinton literally spent the last of his savings to travel to Perth to talk to Mark.

Quinton soon found a job as CEO of a tiny Nevada gold

company named Evolving Gold. He made just enough money to prepare for college expenses for his three children of high school age, and to go pester Mark every six months about doing a deal.

CHAPTER 7
WHERE I FIT IN

I RUN A POPULAR GOLD WEBSITE called 321gold.com. I've done it for almost twenty years now. My beloved wife Barbara was the brain and I was the brawn. Several times a month I would go visit mining projects. She was delighted to have me out of her hair.

Other than having owned and run a couple of small placer projects, I don't have the technical background that many of the other newsletter writers have. That said, I've been to a lot of projects on most of the continents. If you sent a monkey to enough different properties he would develop a pretty good idea of what makes a successful program.

I caught the beginning of the Vietnam War and became a fighter pilot in the Marine Corps at the age of twenty. I spent twenty months in Vietnam, flying over eight hundred missions in fixed wing aircraft and earning a few "I was alive in '65" medals. They made me a captain at twenty-two, and at the time the next youngest captain was twenty-five. I was in the wrong place at the right time. I've written a book about my military experiences with such perspective as I've gained over the last fifty years. You will never find anyone who hates stupid wars more than I do. And they are all stupid.

I did the typical college thing after getting out of the service in 1970 before going to work in computing. Back then computers were bulky, slow, and expensive. My, how things have changed, but the theory is still exactly the same today. By 1974 I had become bored with the pace of writing code, so I found a job delivering small planes all over the world. I made about 240 deliveries, across all of the oceans. I set a dozen international records, won three air races, and even managed to squeeze in a flight under the Eiffel Tower.

One trip in particular is noteworthy. In 1976 I flew a Rockwell 685, previously owned by Wayne Newton, from California to Melbourne. For once I had a passenger: Greg Hayward, the son-in-law of Lang Hancock of iron ore fame. That was a true adventure. We barely made it after a few hair raising emergencies including a engine out landing, a catastrophic oil leak with a full load of fuel on board and ten hours of flying with no battery. Just your average ferry flight.

During the seventy-five hours or so of flight time, as we hopped from one island to another on our way to Australia, I listened to Greg's tales of how rich the Pilbara was in iron, in the form of banded iron formations. Thirty-five years later, that little kernel of information would prove very valuable.

I've seen photographs of Gina Rinehart, the richest person in Australia and the daughter of Lang Hancock, standing next to a painting of her dad. That's my 685 in the background of the painting.

In October of 2008 Barbara got a call asking if I was available to go to Colorado to meet with the president of Evolving Gold, and then to go with him to Wyoming to look at the Rattlesnake alkaline gold project. Since I was just across the border in Nevada, looking at some projects for another company, it was easy to fly to Denver and drive to Longmont and spend the night there.

Barbara's caller from Evolving Gold was Quinton Hennigh. He would pick me up at the motel the next morning and we would be on our way to Wyoming.

He did so. I didn't waste a minute in getting to the point. "I put a bunch of money into Evolving at ninety-five cents, six months ago, on the basis of you guys having the extension to the

Carlin Trend. That's in Nevada. We are now in Colorado, on the way to another project in Wyoming. My shares are now worth fifteen cents. What the fuck happened to the Carlin Trend? I just came from the Carlin Trend and no one wanted to show it to me."

Quinton looked a bit sheepish as he explained. "The financing guys got a little carried away with the Carlin extension story. That's a great place to spend a lot of money and not have much to show for it. We are going to see an alkaline system that I found when I was working at Newmont. We can see a lot more progress there for less money than we could in Nevada."

I was hardly mollified.

The journey from Longmont to Wyoming took about four hours. Over the course of the site visits I have made, I have noticed that on the occasions when I could actually sit and chat with someone for hours, I could learn far more about the project and the character of the company than in any corporate briefing, even those of almost unendurable length. The same was true when I was making a living as a ferry pilot. Most of the time I learned far more in the bar than in a plane.

Mining execs must constantly pitch their company and whatever project is on their agenda. This is because the process of resource exploration consumes money like a school of starving piranha gnawing on a wayward cow.

They might as well record the pitch and play it to you while they go and have a drink. I never learn anything from the programmed pitch because they are telling you want they want you to believe, and are deliberately preventing you from asking questions in order to learn what you really need to know.

Quinton is a little bald and slightly portly; call it pleasantly plump. But then, he lives in Longmont, Colorado. It gets a bit chilly there in wintertime. No doubt wife Heather appreciates a little blubber keeping her warm and snuggly on those bitterly cold nights.

There really isn't all that much you can say about an alkaline system. You have to go see it and do the typical pointing and arm-waving to get any feel for it. So if you happen to be on a four-hour car journey, you have to fill the time with chatter about various other things.

Being in a vehicle alone with Quinton Hennigh for hours was a gift. He is one of those very bright guys who can talk intelligently on a wide variety of subjects. There was his work for Newcrest of Australia, and Newmont in the U.S. After the Bre-X fraud was exposed and the industry shut down for all practical purposes, there was his time as a middle school mathematics teacher. It showed; he was always a great teacher.

We sorted out world peace, and how to fix the financial system. We finally got into a discussion of how the gold got into the Witwatersrand Basin. While working for Newmont in 2005, Quinton wrote a paper about a pet theory of his that he had been working on since his PhD studies.

According to his calculations, 2.8 billion years ago seawater would have had the capacity to absorb gold at a concentration of between four and forty parts per billion (ppb). At a concentration of 10 ppb, one cubic kilometer of seawater could have contained 300,000 ounces of gold. Today, seawater still contains gold, but at a vastly reduced concentration of about 322 ounces per cubic kilometer, or ten parts per trillion.

Where did the rest of the gold go?

He identified a number of basins of similar age around the world. The area with the greatest potential seemed to be the Pilbara Basin, in Australia's WA province.

I was intrigued by the theory. I knew of the two main competing arguments about how the gold got into the Wits, one being that it was some form of modified paleoplacer, and the other that it was hydrothermal in nature. You can take any two geologists in the world, sit a case of beer between them, and they can fight for their preferred candidate until they run out of beer or fall off their stools.

His theory was interesting.

We arrived at Evolving Gold's Rattlesnake deposit. We were just in time to see the drill crew dragging the machine out of the field as snow started to fall. Had we visited one day later we would have been either snowed out or snowed in. The site visit was about as close to useless as any trip I ever made. I saw some dirt and a lot of snow. There was a lot of fog as well. Who knows, there may well have been some gold underneath all that dirt.

Snowfall at the Rattlesnake deposit

So there wasn't much to see. Evolving Gold had made some interesting hits, but as with all big systems, it takes a lot of money to define a large resource.

Quinton's work at Rattlesnake was the success he believed to be possible. The stock went a lot higher as the market recovered from the crash in 2008. But he wanted to follow up on proving the Pilbara gold theory. He just knew it was the next giant gold discovery.

Quinton and I soon became best friends. I had three placer projects, one in British Columbia in Canada and another in Tanzania, and a small operation in Sonora in northern Mexico. Quinton advised me on all of them. In the dozen years since that first meeting we have spent a total of several months together, and have always found something interesting to talk about.

On one occasion, Heather managed to drag him down to Grand Cayman to spend a week with Barbara and me on vacation. I already knew Quinton could find gold, but while swimming thirty meters out from shore he found five one-dollar bills. I can imagine finding one bill, but how on earth do you manage to find five?

He would be even more successful finding gold in the Pilbara if he could only conclude a deal with Mark Creasy. Alas, that's a high hurdle to jump.

CHAPTER 8
PLACER GOLD ON THE SOUTH ISLAND

AFTER THE DUST HAD SETTLED from the first Global Financial Crisis in 2008, I returned to my pattern of making two or three site visits a month. In May of 2009 I got a call from Quinton asking if I could block out three weeks, to make a trip to the South Island of New Zealand for ten days or so, and then to go on to Perth, where he would try once again to make a deal on the Pilbara with Mark Creasy.

It was an unusual trip for me, as until then I had never paid my own way. After all, I was going to write about and bring attention to the companies whose projects I was visiting, so they paid airfares and expenses.

Quinton mentioned that he too would be paying his own airfare, since the purpose of the visits was eventually to do a deal with Mark Creasy. That deal would be with Quinton's own company, not with Evolving Gold. And he needed some help from me in the matter of expenses. I readily agreed to that. Every day spent with Quinton was as good as a semester or two in geology class.

I had delivered a couple of airplanes to New Zealand in the 1970s and 1980s but, other than a short hunting trip to the South Island, I had never visited as a tourist. I suppose that was true of everywhere else, too. While I have traveled the entire world and lived outside the U.S. for years, I never made the time to be a sightseer. It's one of those traps we all fall into. The only time you ever explore where you live is when someone comes to visit and you need to entertain.

New Zealand has always been a favorite of mine. When I first landed there, in the 1970s, it was like visiting a country on a commercial flight and hearing the captain say, "We have just landed in New Zealand. Please set your watches back fifty years." What can be bad about a country with five times more sheep than people?

A few of New Zealand's main inhabitants

Quinton and I flew into Christchurch, on the east coast of the South Island. Someone he knew named John Youngson and his wife or girlfriend, Sue Attwood, picked us up. I was interested in finding a placer project if I could.

I never figured out just what the relationship was between John and Sue, but at the very least they were living together. I liked Sue at once and that never changed. I didn't much care for Youngson and that also never changed. He was a hell of a lot more impressed with himself than I was.

In the southern half of the South Island, a rock type can be found that has tiny quartz veinlets running through it, with particles of gold in the quartz. It is very low grade, but New Zealand is a rain forest with a lot of rain, so the rocks are constantly breaking down and the gold is moving toward every possible body of water. After many travels, visiting different kinds of deposits, I never saw as much gold as I did on the South Island, albeit low grade. But you could go to every river and every stream and pan gold. It was everywhere. It was heaven.

Low-grade quartz stockworks in the rocks

John had a plan for the week. We drove around the entire southern half of the island. He briefed us on all that he knew. Every time we saw water, we stopped and panned. Anyone driving by would have thought we were idiots; grown men on their hands and knees, butts in the air, trying to find a pennyworth of gold. Perhaps we were idiots, but when you catch the gold virus your outlook changes.

For a portion of the trip we were accompanied by Donna Falconer, who had been a master's student financed by Newmont while Quinton was still with the company. She shared his belief that gold precipitated out of salt water in the increasing presence of oxygen. Indeed, many of the attributes of gold precipitation and the bacterial remobilization of gold were (and are) taking place in New Zealand.

Part of the reason for the trip was to prepare me to accept the theory, but I was already halfway aboard. I remembered my hours of discussions about banded iron formations with Greg

Hayward while flying Lang Hancock's 685 to Australia.

John Youngson shoveling under a giant boulder

We stopped at one area that had been a stream until the

water had changed course. There were giant boulders. The water traveled down the river and encountered these big rocks in the way. The water would slow and gold would be trapped under the rock. I took a panful from under it. Sure enough, I found some tiny bits of gold in my pan. But we found the same thing everywhere, including gold trapped in moss.

Bob looking for gold under a boulder

It was easily the most interesting ten days or so I ever spent on a working holiday. Quinton could and did pour information into anyone willing to listen.

With so many sheep in New Zealand, the economy is based around livestock and farming. Gold was discovered in commercial quantities as far back as the 1860s, and attracted tens of thousands of prospective miners. Mining drew people to New Zealand and financed much of its early infrastructure, but it was never as important an industry as it was in Australia, particularly Western Australia.

You can't do much with the land in Australia, while New Zealand has water and a mild climate perfect for sheep rearing. And a long history of great sheep jokes.

Quinton Hennigh in New Zealand

Lord of the Rings was filmed on South Island. As we drove west from Christchurch, we passed through the area where the filming took place. I may be biased, but New Zealand is one of the most beautiful places on earth. And having lots of gold never hurts.

I was well aware of the economic catastrophe on the horizon and wanted my own gold mine. The purpose of our being there was to look for placer projects. I am of the belief that wealth springs from an excess of savings, not from an excess of spending. You cannot spend your way to prosperity. Nor can you borrow your way to success.

Bede Falconer panning for gold on South Island

Our party was comprised Quinton, John Youngson and Sue Attwood, along with Donna Falconer and her son Bede, aged ten. We also had a German stock promoter named Werner Ullman, plus me.

Of all the people on the trip, it was undoubtedly Bede who got the most out of it. What an incredible adventure for a ten-year-old boy. He's probably still telling all his friends about standing in six inches of freezing water in a creek near Queenstown, where he discovered the first tiny flakes of gold in his own gold pan. Every kid should learn how to pan for gold.

Parts of the trip were especially memorable. We stopped at a tiny beach on the west side of the island. The wind was blowing like stink, June being midwinter, and the cold made it through right to the bone.

Under the right conditions of tide and wind, this particular beach would be covered with tiny gold flakes right on the surface. People living in the area would bring snow shovels with a wide blade and scoop up the top layer of sand. They would run that through a sluice box with the right flow of water and collect gold every time it showed up.

I was quite familiar with the concept of gold flowing down rivers and out into the ocean, but less familiar with the concept of the ocean bringing back some of the gold on occasion. That's what also happened on the beach in Nome, Alaska. The gold went out in the river water and came back on the tide. Same thing with this beach in New Zealand.

The few small-scale gold miners in New Zealand are adapted perfectly to the conditions. Their mining equipment is so well matched to the country that even at a time when gold was just under $1,000 U.S. an ounce, they could profitably mine at 0.2 grams of gold per cubic meter. That's right, at $6.50 a cubic meter. (We would need this information later, when trying to determine the profit potential of Egina in Western Australia.)

Tide and winds will cover this beach with gold

The South Island portion of this trip is hardly important to the overall tale, but John Youngson would play a key role in the attempted *coup d'état* nine years later by the Toronto Mafia, when they tried to steamroll Novo. Luckily I had kept my trusty quill sharpened, and intervened at just the right moment.

I did arrange to put in a claim on a project where someone had overspent terribly. They had this giant, floating plant. But they were dumping their tails right into the same pond. Eventually someone realized that all they were doing was processing material they had already processed. As you read on you will find more examples of project planning of similar quality and rigor.

Abandoned gold dredge in New Zealand

There was a wonderful, combined gold mine and tourist attraction for sale. It would have made a decent living for a couple thirty years younger than Barb and me. It needed a bit of investment but could have made an interesting project. The owners could then make money from tourists from both sides of the island, because it was located right in the center of South

Island, and could go pan for gold all day if they wanted a break from working for a living.

A gold tourist attraction

Werner Ullman would end up taking over my mining project a couple of years later and then blowing it sky-high. He never paid what he owed to either Youngson or me.

Such is how the mining industry really works.

CHAPTER 9
SWILLING $1,200-A-BOTTLE WINE

AUSTRALIA IS ONE OF THOSE COUNTRIES where the government never runs out of projects to dump money into. Consequently it is constantly trying to raise more money via taxes, even though the country is one of the most important mining jurisdictions in the world. The government takes a bite out of every ton of iron ore, every gram of gold, and every trainload of coal. And still doesn't have enough.

Australians should be the richest people in the world, even richer than the Swiss, but they are not. It all goes for taxes, including some really stupid taxes.

Like the tax on wine.

In Australia they tax wine every year — the same wine. If the tax on a bottle of red wine is X for one year, and if the vineyard ages it for five years to improve quality, then the tax due is 5X.

So, think of Australian wine as similar to bathtub gin. The winemakers pick the mature grapes, toss them in a tub, stomp on them, pour that into a bottle, slap a label on and sell it immediately. It is loaded with sulfites and every bottle comes with a money-back hangover guarantee. No winemakers age wine in Australia. They make it and flog it to the nearest chump. Because of all of the sulfites, it also makes a great "green" weed killer.

Mark Creasy has a fix for this. The first time I saw him use it, I was convinced he had lost his mind.

Quinton and I finished our New Zealand trip and headed for Perth to meet with Mark. Quinton had arranged for a few people to join us there, mostly financial people. We had Craig Roberts, a mining engineer who worked closely with Cal Evert in

Vancouver, along with David Eaton, chairman of the Baron Group out of Vancouver. If Quinton was to form a company to develop the vast gold resources of the Pilbara, he not only needed to cut a deal with Mark Creasy, he required financing. Quinton had worked with the Baron Group since becoming president of Evolving Gold.

The entire mob went out for dinner with Mark, his chief geologist George Merhi and his consigliere, Steve Lowe, at the Old Brewery on Mounts Bay. Mark brought along a bag containing various bottles of wine.

Mark handed over the bag to the wine steward, who evidently knew him. A few minutes later we all had a fresh glass of red wine in front of us. Mark pulled out a small spray bottle from his pocket and spritzed his wine with a couple of shots. I had never seen anyone do that before.

Most of the wine I ever drank with friends came with a screw-off top and cost $3.99 a gallon. We also believed Ripple to be a deluxe wine, if you bought the right year and were satisfied with Night Train Express for everyday swilling.

I leaned over to Mark so he could hear me speak and asked, "What the fuck was that?"

He looked at me for a long moment, as if I had just fallen off a truck filled with green bananas. Then he said, "I sprayed my wine."

"I know that, I could see you. But what did you spray it with?"

Now he knew I was daft. "Hydrogen peroxide. It offsets the sulfites used as preservative."

I thought about that for a minute. Hydrogen peroxide is water (H_2O) plus an extra oxygen atom per molecule: H_2O_2. He was oxidizing the sulfites in the wine in the same way that light and water oxidize sulfide minerals.

I wasn't real sure just how much sense that made. I figured out the next morning at zero dark early, when I awoke with a splitting headache from the wine. I just had to get a little spray bottle and some hydrogen peroxide.

But we had to be on our way to the Pilbara. Soon we were packed and driving to the airport.

George Merhi worked his magic when he learned that we planned to spend a few days inspecting various sites in the Pilbara. He and his two assistants, Rob and Morgan, made a run to a sporting goods store to buy tents and the supplies we would need. He also raided Mark's wine cellar for a case or two. He sent the lads on their way to Newman to await our arrival.

Main Street, Nullagine

We would spend almost a week out in the field. We flew first to the airport at Newman, picked up our vehicle and drove to our first stop at Nullagine, to look at the Beatons Creek area.

Anyone who doesn't understand the importance of mining to the economy of Western Australia needs only to fly into or out of Perth at any hour of the day. The airport and terminal will be jammed with hundreds of mining people on their way to a two-week work period or returning from a two-week work period.

George Merhi, chief geologist for Mark Creasy

George Merhi knew Mark's projects in the Pilbara better than anyone. He had worked for Mark for years, and did most of the exploration and drilling. Rarely would you find him without a big smile on his face and hair going into all directions. I suspect he may have had a run-in with a light socket, and lost.

Quinton knew George and the Beatons Creek area, having visited a few times. But for Craig Roberts and me it was brand new. Here he is, pointing out where the gold would be found.

Quinton shows us the gold conglomerates

We hadn't been at Beatons Creek very long, looking at all the reefs of conglomerate, when Quinton reached into some old workings from over a hundred years ago and handed me a piece of carbon. At that moment I became a believer in the story. Nothing since then has dashed my belief that Quinton had found exactly what he was looking for.

Bob is handed a piece of carbon

While carbon has an affinity for gold, it's unusual to find plain carbon or graphite with a gold deposit. A key to understanding Quinton's theory of gold precipitation from salt water is the matter of where you find it. In the Witwatersrand, one of the fourteen reefs featured a thin sheet of carbon with tiny particles of gold within. Some geologists believe the carbon seam captured the gold in the same way that moss did on the riverbanks in New Zealand that Bede had panned. But it was the chemical attraction of gold to carbon that put the gold in the carbon leader in South Africa, not friction from the carbon catching the gold. It's a subtle but key difference.

Of course, Craig Roberts and I were like kids in a candy store. There was gold everywhere. You could see where miners a century before had taken hand tools and carved out a tiny cave to dig out a few ounces of gold.

We were in Nullagine in the late days of fall. It was still hot. In summer the area would be hell on earth. A nice cool cave to work in might be just the right thing.

Craig Roberts looking at the conglomerate layer

Everywhere we found a dried-up waterway, we would take a sample and pan it. I won't say it was the richest ground I had ever seen, but you could find gold just about everywhere. That is exactly what Quinton believed would happen when he pitched his theory to Newmont four years before.

Bob preparing to pan scrapings

In hindsight, it's interesting how things worked out and what I saw from the very beginning. We were standing on top of a small hill at Beatons Creek. Quinton was doing the typical geologist's pointing and arm-waving. He pointed at some reefs on the side of a hill perhaps half a mile from us. "Do you see those layers of conglomerate outcropping?" he asked. "Each of them has the potential for gold."

I replied, "That's all very interesting, but a lot of the reefs have eroded away. Where did all that gold go?"

Viewing the layers of conglomerate

Quinton laughed. "Much of it was in the creek beds, and where the streams run during the rainy season. That's where the early miners back in the 1890s made their fortunes. The rest of it all went into the Indian Ocean."

For some strange reason, that comment about the gold all going into the Indian Ocean struck me as important. I wouldn't forget it. Later in our saga, readers will understand why it turned out to be so vital.

Not all of the gold made it all the way to the ocean. Some of it dropped out en route, at Egina and the flats to the west, covered with loose gravel.

We spent about five days in the field. All day we would be out in the field on quads, running around banging on rocks and panning for gold. At night we would return to our campground to cook dinner out of cans. And drink wine at $1,200 a bottle.

Bob goes panning

No kidding.

The first night, George and Quinton discussed the wine they should drink. Should they tipple the $750 wine or the $1,200? I just looked back and forth between them, feeling very stupid.

Clearly it was one of those male inside jokes that you had to be part of the group to understand. They understood it. They were insiders. Craig and I being on the outside, plainly didn't get it. Neither of us wanted to look silly by asking what they were talking about. So we just glanced back and forth between George and Quinton.

That first night they compromised, and we sipped the $900-a-bottle wine. I couldn't tell the difference between that and Ripple. I didn't feel like advertising my ignorance so I pretended I was also an insider and understood it.

We could cover a lot of ground on quads

Next evening my curiosity got the best of me when they were comparing the bouquet of the $1,200 wine to that of the cheap stuff, the $750 wine.

"OK, guys. You have to let me into the club. What the hell are you talking about, twelve-hundred-dollar wine? It couldn't possibly cost that much."

George pointed out the obvious: "That's in Aussie dollars, so it's a lot cheaper than you think. It doesn't cost twelve hundred U.S."

I suppose he thought he was making things easier to understand but it was hardly clear to me.

I responded, "We are in the middle of the Australian outback sleeping in tents in a campground after pounding on rocks all day. What the fuck are we doing, drinking twelve-hundred-dollar wine? Or nine hundred dollars, or seven-fifty, or wine at five bucks a gallon. Have we lost our minds?"

George smiled and said, "Mark gave it to us. Anytime we go out into the field we take a case or two of wine. Some of it is pretty expensive."

That was what you might want to term an understatement.

"Mark seemed *compos mentis* when we were out to dinner. I don't care how rich he is, he won't be rich long, handing out cases of wine at those prices," I said.

George explained, "Mark likes wine a lot. He will come across a vintage he favors, and buy cases of it. Eventually he tires of that stuff so he puts it in his warehouse, and we take it with us on our trips to the field. It's a treat for us."

"I wish someone had said something to me in Perth," I began. "I would have been happy to take the cash instead. I could have handed my camera to you. You could have brought me the pictures back in Perth. I'd love to take the money and run."

Frankly I couldn't taste the difference between the $750 plonk and Night Train Express, except Night Train Express doesn't leave you with the same hangover.

Tent City in the Pilbara

The gold Quinton and George showed us was in a thin layer or reef of conglomerate material. You could look at the surrounding hills and see where the conglomerate reefs came to surface. Quinton was in his element. George had had far more time on the ground but he didn't have the benefit of Quinton's theory.

Even though Quinton had been in the field several times since his first visit in 2005, while with Newmont, he seemed to know the deposit so well that you would have supposed it had taken him years of exploration to get to that point. But he didn't need it. He understood it better than anyone, right back to 1888. But even Quinton would be surprised a few times in the future.

Left to right: Quinton Hennigh, George Merhi, and Bob

The richest gold in conglomerate

We saw the project at Beatons Creek, which had oxide gold of relatively coarse size. Quinton wanted that project. Millennium Minerals was in the process of spending $100 million to construct a mill in Nullagine and would soon be in production, but their material was mostly sulfide and across a fault from Beatons Creek. Quinton saw a lot of potential at Beatons which Millennium failed to see. Over the five days we covered a lot of territory, including a visit to the Millennium mill.

The Millennium Minerals mill at Nullagine

On the last leg of our tour we visited Marble Bar, reputed to be the hottest town in Australia. The original settlers back in 1891 believed the rock formation was marble. Actually it was jasper, but beautiful nevertheless. There were several sites around Marble Bar that Mark Creasy had controlled for years.

Quinton wanted them too.

He dressed up in his finest outfit to take us to see the Marble Bar. The primary thing that I got from this visit, and everywhere else we went to, was the enormous scale of the gold area. It could only be explained by something as simple as Quinton's theory of gold deposition, but everyone else had missed it. There was gold everywhere we went. And it was all connected.

Quinton Hennigh at Marble Bar

On our last night in the field we were staying in a campground near Marble Bar, sipping on a nice mild $750 wine, when a bus pulled in. A seeming herd of young kids flowed off the bus. They looked to be thirteen or fourteen years old, right at that dreadful time when their hormones kick in and at the same time they discover their parents are stupid. Several hundred of them came down the stairs, each clutching a cellphone.

We all laughed at the thought of them, determined to spend

as much time as they could chatting on their phones with their BFFs, who just happened to be sitting three rows behind them. Where did they imagine they would recharge their phones?

Even in 2009, kids needed a phone to travel

Clearly the tour director was wise to the ways of teenagers. The kids got out, set up their tents, and by the time they completed their work the matron in charge had set up a table and a connection to a generator on the bus. I guess they had run into this issue before, and maybe most adults were just as dumb as their parents.

We made our early flight back to Perth from Newman and drove into town to sit down with Mark, to chat about our visit to the Pilbara and possible terms for a deal.

Mark Creasy is one of those people hard to define. He would probably be a great poker player as he keeps his cards close to his vest. He's interesting, intelligent, and rarely misses a trick. Quinton, Craig Roberts and I sat down to meet with Mark and his accountant, Steve, in Mark's office.

I will give him full credit for knowing how to build the ultimate man cave. For twenty years I delivered small aircraft all over the world. In the U.S. at that time it was possible to buy a nice, fairly new light plane for about the cost of a new car. In the rest of the world, with their incredible taxes and duties, for all practical purposes all of the people to whom I delivered aircraft were rich.

Mark is rich, but he spends his money on what is probably the world's finest man cave. We live in strange times, where young people no longer know what sex they are. And it seems there are twenty or more types to choose from. I didn't know that. When my kids were born, we just looked down and knew what we had on our hands right away.

I've never heard of a woman cave. Owning and outfitting a cave seems to be a man thing, not a woman thing or a thing pertaining to any of the other variations of sexual orientation. Mark does a man cave right. He had a large round table in the middle of the room for us to sit around, so everyone could speak and see the others. In one corner of the room was a giant stainless steel tank of some sort. The rest of us, all being male, naturally had to know the story of the strange tank.

It seems that in July of 1979 the first U.S. space station, Skylab,[13] was about to tumble to Earth. The boffins who controlled the vehicle attempted a course change that they thought would bring it down in the Indian Ocean. Alas, parts of it landed in Western Australia. Mark determined the approximate path of the ship and drove out to the desolate area where he supposed it would impact Mother Earth. He located the stainless steel tank and a few smaller pieces of Skylab.

I said that he rarely misses a trick. While he was searching the Fraser Range for those pieces of the ship, he also found indications of nickel.

When he returned to Perth with his new and old treasures he immediately got on the phone to NASA, and asked them how much they would pay for a couple of slightly used oxygen tanks, previously owned by a little old lady and only driven on Sunday. NASA laughed and said he could keep them, with their blessing.

I have cast-iron proof that not only is there a God but also that she has a sense of humor, I offer his subsequent research into the nickel in the Fraser Range. He came across documents showing a joint venture between Newmont, WMC and Anglo American from twenty years before. The JV found intercepts of copper and nickel. At the time it wasn't economic because nickel was still in the discount sales bin, but it later increased in price.

Mark staked a large nickel deposit that became the basis for Sirius Resources' Nova-Bollinger mine. He later sold his 30 percent stake in Sirius for hundreds of millions of dollars. As if he really needed it.

That wasn't quite good enough for him, as he was out the

cost of the gas to search for the Skylab bits, so he did a deal with Legend Mining on another nickel prospect in the Fraser Range [14] and kept 28 percent of the shares. That stake is worth $75 million today.

Mark always wants to be sure of the names of the people he is dealing with. He asked Craig Roberts again what his name was so he could firmly implant it in his mind. Craig told him. Mark looked as if he didn't understand and asked once more.

And pondered again after a pause before saying, "Oh, Craaayyyghhh" with a broad Australian pronunciation. Then they both laughed. While Craig had grown up in Australia, he had lived overseas for many years and had literally forgotten how to pronounce his own name in Australian.

The meeting was mostly a rehash of what Quinton knew from his previous trips, plus some thoughts from Mark. In private, Quinton and Mark would discuss terms, and what Quinton wanted in the way of ground. Mark still wanted someone to write a check with his name and a lot of zeros on it, but in June of 2009 his holdings in the Pilbara were best suited for raising kangaroos.

The Pilbara offered sniffs of gold. Mark had drilled a few holes showing gold, but in 2009 it wasn't a deposit with a known or even a predictable resource. No one was about to write a big check without doing millions or tens of millions of dollars' worth of exploration.

At the time, Quinton knew that he needed Mark but Mark hadn't yet realized just how much he needed Quinton. No man is an island. No kangaroo pasture is a mine. Mark had had it way too easy in the past. But he was under no financial pressure, so Quinton was going to have to drag him kicking and screaming into a deal.

As we concluded the meeting, someone — probably Steve

Lowe, the number cruncher — asked if anyone had any other questions or comments before we concluded. I hadn't contributed much to the discussion prior to this, so I spoke up.

"Well, you have most of what you need to make a giant gold discovery," I began. "You are missing only one key element."

I nodded to Mark. "He's got the ground."

And then towards Quinton. "He's got the interesting theory."

On to Craig: "And Craig will help find the money to move it forward. And you, Steve, are in charge of all the accounting paperwork bullshit."

By now I had their full attention. What on earth was I talking about? What could possibly be missing?

I spoke sagely, as if I knew very well indeed what I was talking about. "You can't have a giant gold discovery without someone to tell the story. If you don't tell the story, there is no story."

Everyone in the room looked at me as if I had lost my mind.

"I'm not doing anything else right now. I'll tell the story, unless we can find someone better." (I didn't mean that last part, of course.)

I got some strange looks. It was something no one else in the room had considered up until that point. They all pondered it and sort of mumbled, "OK."

I still think it needs someone good to tell the tale. How can you have a gold rush without a Mark Twain or Bret Harte or Jack London? I wonder who could write it? We need someone really good.

CHAPTER 10
GALLIARD AND THE EARLY YEARS

QUINTON HENNIGH'S GOAL was to turn the Pilbara into the gold producer that he was convinced it could be. There was this one obstacle. Mark Creasy controlled much of the Basin, and dealing with him was a slow business.

The $117 million Mark collected from Great Central Mines made him rich. When you have that much of it, money tends to lose its ability to motivate. But he never ceased being a prospector at heart. He had, and has, pieces of dozens of Aussie juniors, including several that have multiplied his fortune. He certainly wasn't a one-trick pony. I don't think he has made it to billionaire status yet but he'll get there one day soon.

So Quinton was holding a burning candle, but neither time nor money moved Mark. Quinton presented his case and made an offer in June of 2009. Time passed and the candle became shorter.

Quinton formed Galliard Resources in October of 2009 from a shell controlled by Robert Bick. They raised $400,000 at twenty cents a share to keep the door open and the lights turned on, in anticipation of doing some sort of deal with Creasy. Quinton was eager and ready to follow his dream.

One part of Mark Creasy's deal-making technique is to leave the other party hanging. Quinton had no doubt about the potential of the Pilbara. He had lined up financing, taken a shell company public, and was ready to do the mundane day-to-day exploration necessary to turn scrub ground into a gold mine. Mark wasn't in such a hurry. He knew it cost $1 million a year just to file the financials, keep the stock exchange happy, meet payroll and pay the light bill. But it wasn't costing him $1 million a year so he could take his sweet time in doing a deal, knowing that the other party was under pressure.

Galliard did a second, more serious financing in January of 2011 to raise $1.6 million at twenty-five cents a share. I didn't participate in the first placement but bought as many shares as I could in the second: just under 10 percent of them. The reporting requirements and rules for selling blocks of shares change at that level, so unless you plan on being in the shares for a long time, most investors will chose to stay under 10 percent.

By accident, Quinton then trumped Mark. Quinton had all his ducks lined up in a row before we went to Perth and the Pilbara in June of 2009. He wanted to get moving. His goal wasn't just to make a giant gold discovery; he had higher aims. But first he wanted to find an economic deposit, put it into production, and expand it until he was running the largest and most profitable gold company on earth. But he had to do a deal on that kangaroo pasture before he could do anything else, and Mark proved hard to budge.

So Quinton began talks with Millennium Minerals, which was in the process of building a mill and had dozens of satellite gold deposits near Nullagine. We had been banging rocks at Beatons Creek, which Millennium also controlled. They saw no potential in the project, but Quinton did. He was on the verge of completing a deal with Millennium but also wanted the adjoining ground owned by the Creasy Group. When Mark realized that Quinton was about to do a deal with Millennium on Beatons Creek, he came back to the table and became serious about doing a deal with Galliard.

Galliard and the Creasy Group signed a non-binding memorandum of understanding in February of 2011. It would give Mark Creasy 6.6 million or 43 percent of the shares in Galliard in exchange for 70 percent of his land position at Marble

Bar and Nullagine. It was essentially the deal that Quinton would have signed in June of 2009. Mark had dragged his heels for eighteen months in the hopes of an even better deal for himself.

But Mark wasn't through dragging his heels just yet.

In April of 2011 Galliard completed the deal with Millennium Minerals on Beatons Creek, agreeing to fork over $500,000 worth of shares and to spend $1 million on exploration to earn a 70 percent interest.

All of this time Quinton was the brains and most of the brawn behind Galliard, but he was also still president and CEO of Evolving Gold. From the dismal days of October 2008 at the bottom of the market, when Evolving shares dropped to a low of fifteen cents, Quinton worked his magic at Rattlesnake and increased the value of the shares by 1,100 percent over the next year. He turned Evolving Gold over to William Gee in March of 2011 and left entirely that November.

Quinton is a great consulting geologist but not quite as good at picking his own replacement. Gee took over a good company making solid progress, which had done a joint venture with Agnico Eagle on Rattlesnake in mid-2011. But then Gee got sick, and a fellow named R. Bruce Duncan took over.

Now it may be terribly unfair and indeed unkind of me to point this out, but every person I have ever met, going right back to H. Ross Perot, who used an initial for his first name was a pompous blowhard. Duncan was no different. He took a good company, in bed with one of the top mid-tier gold companies in the world on an excellent project, and he ran it right off a cliff. I didn't have all that much to do with him, but after every conversation with him I would contemplate just how pompous he acted and how dismally the stock performed as a result.

He became CEO in early 2012. The stock price never again

touched the height it had under Quinton, and Evolving Gold finally evolved into bankruptcy.

I have nothing against arrogance, but if you insist on being arrogant you should at least have done something worthwhile in your life. Duncan was a legend only in his own mind. He took a solid company with a good partner and destroyed it.

Quinton took over formally as president, CEO, and a director of Galliard in April of 2011. In June they changed the company name to Novo Resources. I hated the name Galliard; it was like listening to a cat trying to climb a chalkboard. The name just grated.

With the change of name came an increase in the speed of the work being done. Quinton was finally achieving the initial phase of his dream. Creasy remained stubborn, but Novo could forge ahead as a result of doing the deal on Beatons Creek with Millennium. Novo did a private placement in late 2011 and raised over $5.6 million for exploration.

Novo started a 5,000-meter reverse circulation drill program at Beatons Creek in the fall of 2011. By February of 2012 they had completed 80 percent of the program of forty-five vertical holes, ranging from 50 to 250 meters deep.

But Novo began to run into a problem that would later turn critical, and I'm going to illustrate it with a picture.

Half the gold ever mined in all of history has been mined since 1965, when low-grade open pit heap leach deposits as employed in the Carlin Trend in Nevada came into common use. Gold fetched only $35 an ounce then, so the industry needed to find a cheap way of producing it.

Unlike most of the gold taken from the Earth's crust up until that time, the gold in Nevada was microscopic. You can't see anything, even in the richest Carlin-style gold. The rock is butt ugly. However, the tiny size of the gold particles makes it perfect for a heap leach. More mining vocabulary: a heap leach is where the ore is stacked up in big piles and a solution of cyanide is literally dripped through it. That pregnant solution of cyanide loaded with gold is further processed to produce bars of gold (or often with some silver) called doré. Only then does it get sent to a refinery for further processing.

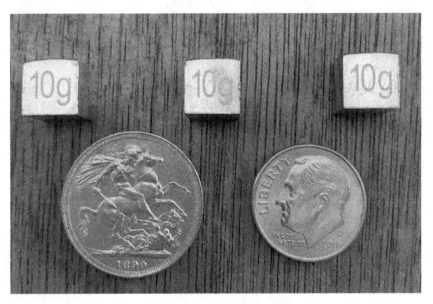

The size of thirty grams of gold

Fine, microscopic gold is desirable for two important reasons. Generally, it tends to be found spread throughout the rock containing it, so drilling the rock gives a consistent indication of grade. Secondly, the smaller the gold particles, the easier it is for the cyanide to dissolve them.

Let's go back to the picture above. I show three ten-gram pure gold cubes, about a quarter of an inch along each side. The British gold sovereign and the U.S. dime are for scale, so readers will understand how tiny those gold cubes are.

Imagine that you possess two cubic meters of rock that you

believe contains gold. If you have a dining room table for six people, you have about two cubic meters of air underneath the table top and between the table legs.

One cubic meter of rock weighs about 2.5 tons, depending on its density.

Now imagine those three gold cubes are located somewhere within your two cubic meters of rock. The grade would be thirty grams of gold divided by five tons of rock, or about six grams per ton. At today's price of gold, a ton of that rock would be worth about $360. So the rock is quite valuable. Any geologist or mining engineer would be happy with six grams of gold from each ton of near-surface rock.

But if you have those three little cubes in a block of two cubic meters, how do you find them? You could drill that rock under your dining table twenty times and not hit even one of the cubes. That's called the nugget effect. It's why geologists always want gold to be microscopic, firstly so they can find it, and secondly so they can process it. You could drop one of those cubes into the cyanide solution used all the time in heap leach deposits, come back six months later, and it still wouldn't be dissolved.

Quinton was aware of the nugget effect. What it meant was that his drill results were never quite as accurate as they would be if the gold was both microscopic and uniformly distributed within the rock being drilled.

The gold at Beatons Creek proved to be coarse from a placer miner's point of view but small enough to give Novo a fairly accurate measure of grade. In what would be one of the very few times I disagreed with Quinton, I took a sample and panned it. I told him that his theoretical grade was off by about 50 percent. He had more gold than his tests indicated. It would take almost two years for me to be proven right.

Quinton has a knack for understanding geology and his

explanations almost always make complex things simple. In this case, his method of measuring gold was inherently inaccurate because of the nugget effect.

Relatively coarse gold from Beatons Creek

Novo and Quinton hit the ground running in the fall of 2011, starting a 5,000-meter drill program at Beatons Creek to earn the agreed 70 percent interest in the property.

Those outside the mining industry probably believe that the hard work in mining is the discovery of the mineral in question. Actually that is just the beginning. Permitting and building a mill might take several years and cost anywhere from $50 million to a billion dollars, or more. And that is after you convince the mines department that you have an economic deposit that is safe to mine. The discovery is the quick and easy part. Millions of dollars will have to be spent on surface exploration, environmental studies, heritage investigations, and lots of drilling.

Actually, Quinton and the Novo crew had it easy. Quinton's theory was that the gold was in a certain sequence of reefs that could be identified by eye. His theory held that there would be gold throughout the reef but not above or below the reef.

When a company drills for a mineral, it is just as important to know where the mineral is not as where it is. Running a ton of rock through a mill costs the same, no matter the economic value of what it contains.

Galliard's (later Novo's) agreement with Millennium required it to hand over $500,000 worth of shares and to spend at least $1 million in Aussie dollars within two years to earn the 70 percent interest. Millennium was busy finishing its $100 million mill and couldn't be bothered with the gold from tiny Beatons Creek. In hindsight, that was a massive error of judgment that would eventually put Millennium into bankruptcy.

The gold from Beatons Creek was particularly suitable for the mill Millennium had designed, while the deposits Millennium planned to mine were low-grade and generally unsuitable for the mill. For all the paperwork and all the testing that government agencies require in Australia, or anywhere else, no one ever insists that a mine plan or mill design must make sense. Quinton and other observers in WA watched in wonder as Millennium thrashed around like a headless chicken for almost ten years, trying to make a silk purse out of a sow's ear before falling over dead.

Junior mining companies rarely go out of business because they have a poor project. What kills more small companies than anything else is making a poor job of managing cash flow. Mining in general, but especially exploration, requires a constant infusion of cash. From the market crash in 2008 until late 2011, the cash spigots opened wide as the price of gold climbed higher and higher, to a new record. Quinton was especially well connected and respected in the tiny mining community. He had little problem raising cash. Others broke their picks and crashed.

The newly-named Novo Resources began releasing assay results in February of 2012. Those numbers certainly justified the deal with Millennium. Almost all of the holes intercepted the reef, with mineralization widths of as little as two meters and as much as eight meters, with an average of about four meters. Four meters is a nice block of material to mine, and it was all connected.

Quinton had drilled within an 800-meter by 800-meter square. If you assume an average of two grams of gold to the ton and an average thickness of four meters, that multiplies to a little over 400,000 ounces of gold. For a 5,000-meter drill program, that is a lot of gold, developed cheaply. For an initial drill program, that would be a home run out of the box.

Quartz veining in drill core

There are two kinds of drills that are most often used in junior resource exploration. When it is necessary to understand the structure of the rocks being drilled, core drill rigs are used. They have a drill head shaped like the end of a straw, lined with industrial diamonds, and will cut through anything. The extracted material maintains its shape and goes up what looks like a tube, all the way to the surface. The drill crew pulls up that piece of drill stem, carefully removes what is called the "core" and places it in boxes identifying the hole it came from, and the depth from which it was lifted. Core drilling provides the geologist with a lot of information, but is expensive.

The other common type of drill rig performs what is called "reverse circulation" or RC drilling. An RC rig grinds up the rock into tiny pieces and blows those pieces back up the drill stem, where they are caught and split into smaller fractions for assay.

Since Novo was drilling conglomerate at Beatons Creek, there was no need to understand the structure — only when they entered and exited the targeted reef. RC drilling doesn't tell the geologist anything about structure, but it's a lot cheaper. When you don't need to know about structure you get a lot more bang for the buck with an RC drill program.

Quinton signed the memorandum of understanding with Mark Creasy in February of 2011. It then took until July of 2012 to drag Mark across the finish line to sign an agreement. He was used to calling all the shots when he did a deal. One of his most effective techniques was simply to delay and delay until the guy on the other side of the table finally gave up and bent over.

By doing the deal with Millennium and delivering an extremely successful drill program right out of the chute, Quinton proved that he was serious, and showed that he was willing to work with somebody else if Mark wouldn't belly up to the bar and sign that piece of paper.

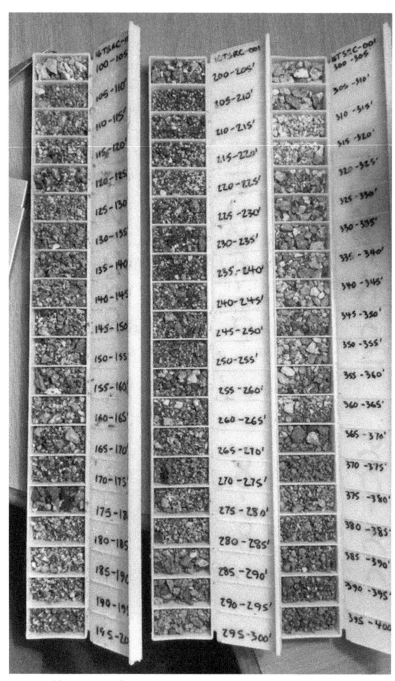

Chip trays from reverse circulation (RC) drilling

Quinton managed to put together an impressive land position in the part of the Pilbara that he thought offered the most potential. On the other hand, Mark Creasy's side of the deal meant that he was now by far the biggest shareholder in Novo. Keep that fact in mind, because two years later it would become a giant issue.

Both were happy for now. Quinton had his teeth solidly into a proven gold property. Mark in effect had Quinton doing all the heavy lifting for him.

Quinton continued work at Beatons Creek into late 2012, with a second drill program of another 107 holes. Results continued to be excellent.

I waited for three years to write my first piece about all this for 321gold.com. It took that long to get Mark to agree on terms reasonable to Quinton. While Novo's deal with Millennium proved to be the catalyst to move Mark off of top dead center, or TDC,[15] it was the ground that the Creasy Group controlled that would turn Novo from just another Canadian gold junior with an interesting project into a gold powerhouse. Quinton wasn't seeking just one medium-sized deposit. His theory was that the Pilbara had the potential for millions or tens of millions of ounces of gold.

My first of dozens of articles on Novo Resources appeared on the website on August 15, 2012. I titled it "698.3 Square Miles of the Wits"[16] since most of my readers are American and hopeless at anything as complicated as the metric scale. I include myself among them.

In the article I explained the basic geology, and Quinton's theory, and the proof we had put together so far. I talked about banded iron formations, since they seemed to me to be such an important clue, although ignored by the geological world.

At the time the company had a $14 million market cap and

an enterprise value of $5 million. I went on to say, "I rarely give a pure buy signal. This is a buy signal, the most important that I have ever made. The stock at $.44 is like stealing." A couple of paragraphs above that I pointed out that investors were getting nearly seven hundred square miles of the Wits. "It is easily a ten-bagger. It could be a 100-bagger. It's going to be big."

So far it has gone up only twenty-fold. But give it time; Quinton has big plans still.

About the same time, in mid-August of 2012, Quinton had Novo put some money into a private placement with Evolving Gold.[17] He knew the potential of its ground and, should it advance the work he had begun, he wanted Novo to have a piece of it. Novo bought two million shares at thirty cents. People in the industry vary in their opinions about just how wise that was.

The other company I visited in Nevada just before flying to Colorado to see Quinton in October of 2008 had about $12 million in cash at the very bottom of the resource mining stock crash. Since it had partners funding all its projects, I suggested putting some of that money into juniors that were being given away. A year later most of those stocks were up a thousand percent or more. The company ignored me and ended up blowing all its cash. That's how the mining business really works.

I happen to agree with Quinton's theory. And over the years he had Novo invest in half a dozen projects and companies. Some turned into giant home runs. Some, like Evolving Gold, turned into a duster.

Novo released the last of the initial 5,000-meter drill program results on August 21, 2012.[18] The company had drilled forty-three holes and found significant gold in forty-two of them. In the two weeks after I wrote my first piece, the share price shot up from forty-five cents to eighty. While more and more juniors pulled in their horns in preparation for a long cold winter as gold prices dropped, Quinton continued to charge ahead.

The holes were interesting; many had a short interval of maybe one or two meters of high-grade gold within a longer, low-grade interval. Quinton always planned for production so he needed to know the distribution as well as the size of the gold.

Again, most junior resource companies don't fail because of their project or projects, but rather from poor management of cash. Drilling eats money. The money is there to make progress with, but investors hate paying big checks to support the lifestyles of the stock promoters and management.

Quinton was one of a handful of geologists (the others include Peter Megaw, David Lowell, and Keith Barron) who everyone in the industry knew and trusted. There's a balance to be found between issuing too many shares, to the point where an investor couldn't make money if the company drilled into Fort Knox, and running out of cash but having the world's tightest share structure.

As long as I have known Quinton, he has been able to tread that fine line between too many shares and too little cash. Investors like results. No, investors *demand* results, and it's like going into a Chinese laundry in San Francisco: "No tickee, no laundry."

Quinton was on a roll. He had the ground to be tested. He had money in the bank. The first drill program was followed up in September of 2012 with a 7,500-meter program. There was another two thousand meters of drilling at Golden Crown Hill, where historic drill tests had found gold-bearing conglomerate beds less than fifty meters from surface. Then at the end of 2012 Novo conducted another two thousand meters of RC drilling on the new Creasy ground south of Grants Hill.

Frankly, most of what goes on in the mining business is important but boring beyond belief. Quinton was drilling conglomerate beds that you could see outcropping on the hillside. Ninety-five percent of the holes reported some gold. At times the results were incredible but, to the average investor, boring. The first red meat as opposed to the far more common tofu came in May of 2013, when Novo announced its initial 43-101 resource.

I'll backtrack here for just a few lines. As a result of the Bre-X fraud, the Canadian government prepared and issued a series of regulations that it calls 43-101.[19] The full name is National Instrument 43-101 Standards of Disclosure for Mineral Projects within Canada. It's a set of rules about what you can and cannot say about a deposit. While the government functionaries love their 43-101, I'm not a big fan. It's more petty worthless rules, often ignored. When someone really does bust the regulations they are given a small and meaningless fine for their misadventure.

Novo's press release talking about the new resource estimate was interesting because it allowed investors some insight into Quinton's thinking.

The initial resource was only 421,000 ounces of gold at a grade of 1.47 grams of gold per tonne. That number was derived from 16,100 meters of reverse circulation drilling. Novo also drilled eight core holes, and could use the core to determine average specific gravity for their 43-101 report.

Quinton also did something interesting that I hadn't seen done before. I found it extremely valuable. He mentioned the cost of the drilling. It had cost about $2.5 million for 16,100 meters. So, according to Quinton's numbers, it had cost about $6 to find one ounce of gold. That's an excellent price of discovery.

A couple of years later, when Novo did a bulk sample, Quinton found that the drill results had downplayed the actual results. He was in fact finding gold for about $4 an ounce, and that is nothing short of brilliant.

Gold and gold shares reached their peak of popularity in 2011 after a dozen consecutive years of higher prices. The shares peaked months before the gold price touched $1,922 in September. For the next four brutal years, investors got beat up if they had not been sharp enough to take some money off the table. Gold and the mining shares hit bottom in January of 2016.

During that dark era for investors, few writers were interested in covering new stories. Jay Taylor of *Turning Hard Times into Good Times*[20] was an exception; he recommended Novo for the first time on August 9, 2013. More about Jay in a moment.

I had visited Novo several times by now and was fully conversant with both the theory and the story. I had been part of the Pilbara story since a year or so before the company was born, so it's natural that I covered it early on. But even with my half-dozen or so pieces there on the website for anyone to look at, no other writers took any notice for a year after I began telling the story.

I had almost five years of thinking about the theory and the ramifications, so that gave me an easy advantage. But how many writers are ever in a position to talk about a Witwatersrand lookalike?

Jay Taylor hadn't visited the project, but in his own way he did a great job of telling the tale. I repeat that Quinton excels in explaining complex matters simply. Jay not only understood the theory; he glommed onto a vital issue.

In his piece he said:

"Without speaking to management about this issue, the fact that management mentioned the fact that Millennium is operating a mine and mill may be of importance in the longer run. Indeed, Hennigh told the *Northern Miner* in October of 2012, "If we can get away from having to build a mill the permitting for a smaller operation is really straightforward and could be very quick.""

Quinton had known about the characteristics of the gold at Beatons Creek ever since his first visit, when he was a Newmont employee. It's coarse and hardly suitable for a heap leach. It takes time for cyanide to dissolve gold, and the bigger the gold particles, the more time it takes.

Jay understood that. His story continued:

"Hennigh may have meant to imply the possibility of an open-pit heap leaching operation and thus no need to build a mill or the possibility of working a milling deal with Millennium Minerals. But for now, that's getting ahead of the story. First we need to know more about the size and grade of the deposit and a host of other issues before we think in those terms. Still, the fact that Millennium's milling operation was mentioned in a press release can't help but get you thinking along those lines."

Jay Taylor nailed it with those comments, even if it had taken several years to arrive at that point.

I've been on dozens of trips to various projects with Quinton. He thinks further ahead than anyone else I know in the mining business. Five minutes after he found the first nugget at Beatons

Creek, he would have been pondering just how to process it cheaply. When he learned that Millennium Minerals was in the process of building a mill, he probably thought he had died and gone to mining heaven.

In September of 2013, Newmont Mining surprised both Novo and the market with its announcement of the purchase from two large shareholders of 17.76 million shares of Novo at a price just above market.[21] That would give Newmont control of 35.7 percent of the shares, if it were to exercise all of its warrants.

Newmont was well aware of the potential of the Pilbara through Quinton's work years before. Some self-dealing on the part of its Perth staff queered the agreement with Mark Creasy and left a bad taste in Creasy's mouth. Newmont figured that if you couldn't get in the front door, perhaps the back door was open.

With the exercise of about eight million warrants by September of 2013, Novo brought in $4.8 million, bringing its cash on hand up to $8.26 million. It had about fifty-eight million shares outstanding and a market capitalization of about $48 million at the time of Newmont's announcement.

Juniors that make deals with majors are often left at their mercy. Making a deal with a big company with lots of expertise and money sounds attractive, but the majors enter into deals for their own benefit, not that of the junior. It's OK to get into bed with the pretty lady but it's also important to not catch the clap.

To give Quinton full credit, for as long as I have known him, he has always tried to structure deals such that no single party could take control. On the Beatons Creek ground, Novo had a 70 percent interest and Millennium Minerals had 30 percent. On the ground around Beatons Creek, controlled by Mark Creasy, Novo had the same 70/30 split. Remember that it was Novo's side deal with Millennium that prompted Mark to conclude his deal with

Novo, after dragging his heels for years.

For Quinton, Newmont would be an especially valuable major to deal with. First of all, he had worked for them. There was mutual respect. Second, majors have large staffs and have all the various technical areas covered. If you know what a major is especially good at, having a big company as a partner can be a marriage made in heaven.

Newmont had a strong group doing bulk leach extractable gold (BLEG) sampling.[22] That was ideal for Novo. Novo had a lot of ground with potential; the Creasy portion alone was about 1,800 square kilometers, or 700 square miles. They needed a cheap, fast way to determine where to concentrate their efforts. BLEG is interesting because it determines only the presence of gold in a location, not its grade.

A BLEG survey team will go out and take fine material samples of decent size from streams, both flowing and dried-up. When you have thousands of samples, you can get an idea of where to look for gold, and the distribution. One big problem for juniors with large land positions is deciding where to expend their energy, and more importantly, where to spend their limited money.

The Newmont–Novo partnership would prove to be valuable to both partners. Newmont's large share position would become a problem for them later, however, when someone made an agreement to sell just before very valuable news was released. But that is a tale for the future.

In late 2013 the Department of Mines and Petroleum of WA notified Novo that it had been approved for two grants, each of $200,000 Australian, for the purpose of drilling deep holes further out in the Nullagine embayment.[23] The plan was for one deep hole at Beatons Creek and another at what Novo called Contact Creek, part of its Marble Bar project.

Gold and silver began a perfectly natural and normal price correction in 2011, such as happens in every market. It took gold bugs by surprise. Quinton knows that no markets ever go straight up. He had always sought to keep a healthy cash position in the bank. Investors tend to want instant gratification and scream for companies to do things faster. But often, slower works better.

In March of 2014 Novo concluded another deal with Mark Creasy on an additional 18,000 square kilometers of new tenements.[24] That's another 7,000 square miles to cover. BLEG sampling would be an important technique to use, to narrow down the search for the mother lode. Novo would be the manager, and this venture had the same 70/30 split that Novo had concluded before. Remember, all of this could have taken place four years earlier, but for Mark being dilatory.

Newmont had completed a baseline BLEG survey over Novo's ground at Beatons Creek as early as November of 2013, shortly after its purchase of Novo shares. In March of 2014, after Novo had added the additional 18,000 square kilometers of ground, Novo and Newmont realized it was time to get serious about carrying out a BLEG survey over the entire new property.[25] That began in March and was scheduled for completion by the end of winter, in July. The BLEG process belongs to Newmont, and its team of technical people is the very best in the industry.

The terms of Novo's deal with Millennium called for the completion of a bankable feasibility study (BFS) by August of 2016, in order for Novo to complete its earn-in of its 70 percent interest in the Beatons Creek project. Quinton had seen enough disasters brought on by missed milestones, and didn't intend to be mousetrapped by the terms of the deal. Under the "shit happens" rule, everything that can go wrong will go wrong.

Novo planned to complete all the requirements of the BFS as early as July of 2014. [26]

In August of 2014 the BLEG work was completed and the results obtained. [27] The results around Beatons Creek and Marble Bar demonstrated that Novo had correctly identified where the gold was coming from. Novo immediately began an aggressive shallow drill program at Beatons Creek to expand the near-surface oxidized gold resource.

Millennium Minerals was now in production but had done a rotten job of designing the mill. There would eventually be a resource of a couple of million ounces of sulfide gold, but the mill was designed for oxide ore. Once Millennium ran out of near-surface oxide material, they were screwed. Their dismal recovery of the sulfide rock meant they were really just running a hiring facility, not a mining company. They produced about a hundred thousand ounces of gold a year but never made money. They existed to provide employment and to repay their bank loans. And to allow management to collect paychecks they otherwise would have had to work to get.

But an escape route was at hand. Millennium's partner in Beatons Creek had five years' worth of ore that would be perfect for their mill. Novo had done all the work of locating and defining the conglomerate gold. All Millennium had to do was to agree some kind of arrangement with Novo. They could buy out Novo's interest. Or they could perform toll milling for Novo, whereby Novo mined the gold and Millennium processed it for a fee. Or they could sell the deposit and the mill to Novo and retain a net smelter royalty.

The only really stupid, head-up-the-butt thing they could possibly do would be to do nothing.

So they did nothing.

It seems clear that Millennium Minerals never saw the value

in Beatons Creek, even after Novo proved a nice five-year resource. Since Millennium saw no value in the project when doing the deal with Novo, they weren't about to rethink their position, no matter what new facts emerged. Millennium was convinced that Novo had bitten off more than it could chew, and would eventually walk away from Beatons Creek.

Quinton Hennigh, on the other hand, could not come to grips with the fact that he couldn't interest Millennium in any sort of deal. He didn't need Beatons Creek; he had nearly twenty thousand square kilometers of other ground where he could go find more gold. He knew it was there. He had perfected a way to locate the best areas and he knew how to mine it. And in the worst case, he could define enough gold to justify building a new mill for Novo's use alone.

What baffled him was a company so stupid that it would risk bankruptcy rather than change a business model that clearly didn't work.

Over the years he came to realize that there were no average mining companies in Australia. There were some of the best-managed and technically most astute mining companies in the world, such as Newcrest. He had worked there and knew they were brilliant. There were also dozens of tiny, poorly-financed juniors run by idiots who would have been on welfare and living under a bridge, if they'd ever had to compete on their own merits. In short, Australia has a few great companies and a number of others run seemingly by dodo birds with nothing better to do.

Millennium was in the dodo-run category. Alas, the Australian dollar kept going down as the price of gold slowly went higher, bailing out the idiots who should have been in bankruptcy court wearing nothing but an old pair of jeans with patches on the knees.

By the end of 2014 Novo had drilled off more of the oxide resource and had completed recovery testing, using both a gravity process and cyanide leaching. Tests showed as much as 91.7 percent recovery via a gravity circuit, with an additional 7.6 percent recovery using cyanide leaching, for a total of 99.3 percent recovery of the gold.[28] That's both brilliant and rare.

CHAPTER 11
2014 BRINGS PRODUCTION OPPORTUNITY

IN LIFE AS WELL AS IN MINING, serendipity often plays an important part. A golden opportunity dropped into Quinton's hands in February of 2014, only to slip away three months later.

As I mentioned, Mark Creasy made his first fortune selling the Jundee gold mine to Great Central Mines in 1995. In May of 2014, Newmont Mining announced its intention to sell Jundee to Northern Star Resources for $82.5 million. [29]

You can think of this as tossing a hungry cat into a pen of chickens. It was an amazing episode, complete with self-dealing, Novo's second-biggest shareholder screwing its second-biggest shareholder (that is not a misprint), and a lot of simple old skullduggery.

I keep saying that Quinton was quite close to Newmont. He had worked there only a few years earlier, and now Newmont was conducting the BLEG survey on Novo's tenements, after having injected a fair bit of cash into Novo.

Newmont (and my readers) may or may not have remembered that the deal between Great Central and Creasy included what is called an ROFR; a right of first refusal. That ROFR would come into play in the event that the mine was put up for sale, as distinct from the company that owned it being sold. (And Jundee had been part of a series of corporate moves since 1995, operating under a number of flags.)

If the Jundee mine itself was put up for sale, Mark Creasy had the right to match any offer made, and his bid would be superior to any other. He hadn't forgotten this, and neither had Quinton, even if Newmont had forgotten or had never realized it.

Mark had given that ROFR to Novo in February of 2014 because he recognized it had value to Novo. No conditions were attached to the transfer; he just handed it over. Novo now had the right to match the Northern Star's offer to Newmont, and Novo's money would be deemed better than Northern Star's.

Jundee began as an open pit mine. Then, when the strip ratio got too high, it was converted into an underground mine. Production varied, but was usually between 250,000 and 400,000 ounces of gold a year.

The deal would catapult Novo into the status of a mid-tier mining company literally overnight. Quinton was standing at the goal line. All he had to do to complete his life's dream was to step across the line.

Any time life hands you everything you wanted and it looks just like a beautiful red rose, just be careful and avoid the pricks because they are always there.

An interesting book came out in January of 1971 called *Games People Play*.[30] The author, Eric Berne, discusses the three main games that people play with each other, often without knowing or understanding the potential dangers.

One very common game is the negative sum game. In this game, one or both of the players want to make sure the other loses. Divorce proceedings in the United States frequently turn into negative sum games, because lawyers know that the best way to maximize billable time is to stir up the pot. If they can enrage both sides to the extent that each starts to focus on destroying the other, the lawyers will make a lot more money.

Wives often enjoy this game, because it is an opportunity to right every real or perceived wrong ever done to them. They are quite prepared to beat the soon-to-be ex-husband to death, even using their children as weapons. The lawyers benefit, the husband and wife both suffer financially, and their children wonder why and how they lost a parent. And two people who once promised before God to honor and obey each other until death did them part now despise each other.

A better game is one in which each party realizes it is in its own best interests not to screw the other party. Divorces don't have to be acrimonious. It's pretty stupid, especially when there are young children. Likewise, if you are planning to enter a business arrangement or a marriage, why would you want to beat up your prospective partner? In a positive sum game, each player makes sure the other side benefits. They realize that working together is far more productive than fighting over pennies.

A perhaps more common game is the zero sum game. Here, one player wants only for the other player to lose. Ross Perot, Steve Jobs, and many businessmen actually believe that making sure your partner loses means you win. In a zero sum game you have one winner and one loser.

Mark Creasy always plays zero sum games. He is totally focused on making sure the other side loses in any business deal. In this deal with Newmont and Jundee, that created a giant problem.

Mark handed the ROFR to Novo when rumors began to fly in Perth about Newmont selling Jundee. Frankly, he wasn't in a position to do anything with the ROFR, while it fit Quinton's plan for the future perfectly.

Quinton was familiar with the geology at Jundee. Instead of the million or so ounces of gold that Newmont believed it had, Quinton believed Jundee had a good five to ten years of production potential. He got on the phone and rapidly amassed commitments for the money. After all, if Northern Star could raise the money, why couldn't Novo?

That's when it began to get interesting. Mark Creasy's accountant, Steve Lowe, was part of Chalice Gold. Chalice had money in the bank and was certainly a candidate to take over the mine. So, all of a sudden, Mark Creasy tossed a monkey wrench into the works. Instead of the easy and smooth road into production for Novo — which would certainly benefit Newmont as Novo's largest shareholder, and Mark Creasy as the second-largest shareholder — the deal was blowing up.

Steve would have known how to push Mark's buttons. I think he did. All he would have had to accomplish was to convince Mark that he would get a better deal with Chalice. And I have a sneaking suspicion that some of the Newmont people in Perth may have been playing both sides against the middle just to stay employed.

Newmont had a major stake in Novo, so why wouldn't Newmont want Novo to go into production at Jundee? Just because Jundee no longer fit its plans didn't mean Novo wasn't a better home. I find it curious that a number of former Newmont people from the Perth office found a new home at Northern Star.

Creasy demanded Novo return the ROFR to him. He was going to cut Novo out of the picture and do a deal either with Chalice or with Northern Star.

At the time this took place, I was driving an early Ford Mustang from California to Texas. As I sat in a motel room in some podunk town, Quinton filled me in.

"Mark wants the ROFR back. He's gotten Northern Star to agree to hand him ten million dollars in shares of the company."

"Quinton, I love you like a brother, but in this case you need to tell Mark to go fuck himself," I said, and added, "He's a big Novo shareholder. If he screws Novo he is also screwing himself. Why would he screw himself?"

The answer, of course, is exactly the same as why a dog goes into the center of the road to lick his dick. Because he can.

Quinton is flawed. Not in a major way, as is the case with most of us, but nevertheless he is flawed. He is much too nice a person. At times that gets in the way of doing sensible and reasonable business. I don't have that problem.

Quinton gave the ROFR back to Mark. Mark got $10 million in shares of Northern Star. By Mark's logic, he did well. His $10 million in Northern Star is now worth almost $100 million. He didn't make that much, because over the years he sold off his shares. But he did well.

But Novo would have been producing at least 300,000 ounces of gold a year, and today would have only one-quarter of the number of outstanding shares it actually has. The guy who got screwed the most in the deal was Novo's second-biggest shareholder. Mark cost himself $500 million in gains. I think the term for that is "penny wise, pound foolish."

It was a fleeting opportunity to leap forward. There are always lots of people doing everything in their power to queer any proposed mining deal, provided it brings them short-term benefit.

Overall, Novo had done quite well in the thirty months since it went public. The deal with Millennium had worked out well and looked to have potential for a couple of million ounces of gold. Two deals with Mark Creasy had been signed. Newmont had come on board with some money and all its technical expertise. Novo had assembled a great team and a series of projects in the Eastern Pilbara that had good potential. It wasn't going to be some giant area of twenty million ounces, but what Quinton had put together was well worth having.

But there was this one clog in the machinery that hampered Novo for years. Millennium had spent well over $100 million and several years building its mill. And while everyone else in

Australia and the world understood it was the wrong mill for the feed, Millennium's management (to use the term very loosely indeed) sailed blissfully along, ignoring the fact that it had turned the company into a sort of home for unwed mothers.

There wasn't a chance of Millennium ever making money processing the rock they owned in the mill they owned. In a masterpiece of self-delusion, they also convinced themselves that Novo was stupid and they were smart. They refused to even consider the one solution that would turn them into a mining company instead of an employment service.

Quinton Hennigh and Novo were caught in a trap of their own making. Had they made a similar discovery a couple of hundred miles to the west, they could have taken the conventional route of defining a resource, raising money, and building a mill to go into production. Quinton never made any bones about it. He wasn't running a lifestyle company that existed to pay management large salaries for doing little until the day that shareholders woke up and booted them. He wanted production.

One reliable rule of mining management is not to spend a few years and $100 million building a mill right next to another $100 million mill that desperately needs, and is suitable for, your sort of ore. That would be mad. In short, by succeeding, Quinton was failing.

The shareholders didn't mind much. Novo has always been one of those resource companies with great liquidity. For some reason the theory of gold precipitation seemed popular. You could always buy and sell its shares. That's a lot more rare than you would think, with the penny mining stocks. For years, there was a 200 percent or greater change in the value of Novo shares every year. The stock would double or triple, dribble back down to where it started, and then repeat. To a wise investor who recognized the pattern, a 100 percent annual return was possible nearly every year no matter if you were a bull or a bear.

CHAPTER 12
WHEN ACADEMICS LOOT AND PLUNDER

IF BY NOW THE READER IS NOT CONVINCED that mining mostly consists of lying, cheating and stealing on the part of many of the participants, then I as author have failed. However, the chicanery in the mining business does not compare to the sheer ruthlessness that exists in academia. Professors tend to make pirates look like pussycats.

Whenever someone comes up with a new theory, perhaps the best measure of it actually being true or working is when no one, including the "experts," agrees with it. In fact, the more they make fun of the theory, the greater the chance of it being correct.

In 1982 two Australian doctors came up with a theory that a bacterium caused most cases of ulcers and stomach cancer. After years of study and gathering proof, they presented their case to an audience of surgeons in Sydney at a medical seminar. Literally, they were laughed off the stage. You see, the surgeons made money by cutting into people due to the belief that ulcers were some sort of lifestyle disease. If you could cure a case of ulcers with a simple course of antibiotics, they would lose patients. And, more importantly, income.

In 1985 one of the doctors, Barry Marshall, had himself tested to prove he was free of the bacterium before deliberately infecting himself. Soon he tested positive for the Helicobacter pylori bacterium and came down with a mild case of gastritis. A short course of antibiotics caused both the gastritis and the bacteria to disappear.

Doctors Barry Marshall and Robin Warren eventually won the Nobel Prize for Medicine in 2005, twenty-three years after their discovery. Once everyone else in academia realized the theory might in fact be correct, they rushed to jump onto the bandwagon. By 2005, some twenty-five thousand articles had been penned on the subject of bacteria causing ulcers. No doubt many of the authors claimed to be the first to come up with the concept.

Professor Hartwig Frimmel from the University of Würzburg crossed paths with Quinton in May of 2005, at a meeting of the Geological Society of Nevada, in Reno. Quinton had been assigned by Newmont to form a team to find volcanogenic massive sulfide (VMS) deposits.[31] Think of an undersea black smoker; that's a VMS deposit. They can be very rich in valuable minerals.

By this time Quinton had developed his theory on gold precipitating out of solution from seawater. He showed his paper to Frimmel, who found it faintly interesting but said it was basically rubbish.

In May of 2014, Frimmel published a paper[32] setting out his new and brilliant and original theory of the deposition of gold from seawater through precipitation in the presence of carbon, as the oxygen content in water increases through the action of single-cell bacterial colonies, the first life on Earth. His theory was Quinton's theory, from nine years before, that he had mocked.

When Quinton formed Galliard and began moving forward in the Pilbara, I wrote and published on 321gold.com a dozen or so pieces, talking about the theory and the Pilbara. As the market accepted the idea, Frimmel began to pass it off as his work.

When Frimmel's paper was published, Quinton contacted him and mentioned that it was his idea and work, and that Frimmel had mocked it years before but was now passing it off as his own. Frimmel tossed Quinton a bone and agreed to share credit as authors on a paper to be submitted to an academic journal, *Mineralium Deposita*.[33]

The full name for *Mineralium Deposita* is: the International Journal for Geology, Mineralogy and Geochemistry of Mineral Deposits. That sort of rolls off your tongue.

That paper was published in January of 2015, with Frimmel as the primary author and Quinton in attendance. They called it "First whiffs of atmospheric oxygen triggered onset of crustal gold cycle."[34] It won the best paper of 2015 for the organization.

At the time, while Quinton was a bit miffed, I thought it had its amusing side. After all, I was writing more about Quinton's theory than he was. While professional journals are far more inclined to publish papers written by academics talking about interesting theories than by the guys in the field actually doing the work, Quinton had taken the theory and was putting it to work to produce gold.

It was humorous right up until the time that two minor newsletter writers pulled the same trick on me, two years later. Then it wasn't even a bit funny.

CHAPTER 13
COMING UP WITH A NEW WAY TO MINE

QUINTON HENNIGH'S PRIMARY STRENGTH is his uncanny ability to think laterally; to view things from a different perspective.

Most of the mining activity in the Pilbara, all the way back to the nineteenth century, revolved around hard rock, vein-style mining. That continues even today. Few companies seem to care about the conglomerates or to be able to figure out how to mine at a profit.

The essence of where and how the gold was found in the Pilbara has been known for a hundred and fifteen years. Government documents as far back as 1906 make references to gold in various layers of the conglomerate, just like the Witwatersrand. Vein deposits present problems of their own, such as trying to find them, but the conglomerates were everywhere. Yet were ignored by most of the mining companies.

It has climate issues as well as geology: hot, dry, and nasty, with either too little water or far too much; and wildlife, with the top ten deadliest creatures on earth. Australia has snails and spiders that can kill. The constant barrage from flies is enough to make you consider suicide just to rid yourself of them.

And don't even think about summers in the Pilbara, at forty degrees Celsius (or a hundred and four Fahrenheit) in the shade. Not that there is any shade.

The conglomerate gold at Beatons Creek was in nugget form, which made it difficult to measure accurately but also meant it would be easy to process. However, it was located in the matrix, between cobbles and boulders. You needed the matrix for the gold but not the rest.

If Quinton decided to use the typical mining method of drilling and blasting before moving the material to the mill for processing, 90 percent of what he moved would be waste material. Further, in doing so he would shake a lot of the free gold to the bottom, where it couldn't be removed by heavy equipment.

I have worked in two very different fields where the people around me were often very smart. One was computer operations and programming in the early 1970s, and the other, obviously, geology and mining for the last twenty years.

Computer programmers are often very bright. They are also commonly self-taught. I have never taken any programming training but I was a systems programmer. Everything I knew, I learned on the job. Steve Jobs had a smattering of college, as did Bill Gates. I'm not sure even today that college is any real advantage in learning programming. The industry and languages change far too quickly. What is popular today will be old hat six months from now.

Geology and mining, on the other hand, require years of study. Nearly everyone in mining or geology has at least a bachelor's degree. Master's degrees are common and doctorates not uncommon. I have met more overeducated people in mining than in any other occupation I have ever been associated with.

The flaw in this emphasis on education is that people tend to become too specialized. They are very good in some narrow skill but lack general knowledge. In the course of the many trips I have made, I have often noticed (and often commented) that most geologists believe mining to be about spending money rather than making it. They would all do far better if, while earning their doctorates, they spent a semester working in a 7-11, learning how to sell a quart of milk at a profit. Profitability seems to be the last thing that most mining folks consider in a project.

Here's an example. My first visit to an underground mine was in 2002; it was the Ken Snyder gold mine in Nevada.[35] At the time it was one of the highest-grade gold mines in the world. Our escort was the senior geologist on the project. He had been there for several years, as I recall. We went down to the working face and saw just how narrow the vein was. Their working width was two meters, but it was ultra-high grade. In the week before we were there, they had mined 28 ounce gold measured across the mining width, but the vein was less than 0.5 meters.

We were suitably impressed, picked up all the samples we could carry and returned to the surface. As we exited the portal, one of the skip loaders followed us and proceeded to dump his load on a stockpile just outside the portal entrance. I turned to our guide with a couple of questions.

"You find some extraordinary high-grade gold now and again. The average is much lower, though still pretty rich. Do you need to blend the ore? Or how do you handle the high-grade stuff to get the best recovery rate?" I asked.

"I don't know how they work with that." He pointed up the hill to where the mill was situated. "They do all the processing there. I've never been in the mill."

I was nonplussed. How could you work on a mining project for years, in a senior capacity, and not have some idea of how things worked? How could you not want to see how ultra-high grade gold was processed?

That's not Quinton. He's the most inquisitive bugger I have ever met. He wants to know everything. He thinks about everything. He has come up with a practical solution to a problem before anyone else has even figured out there might be a problem.

Mines rarely run out of ore. They just get mined to a point

where, for one reason or another, the company running the project cannot make a profit any more. So they shut it down and walk away. Likewise, it's common for a project to be bought by half a dozen companies, consecutively, before someone works out how to run it at a profit. There's that gruesome word again, that keeps popping up.

Political risk, commodity prices, and labor and energy costs are all factors as well, but plain old human stupidity is more important than any of them. Ten thousand prospectors and miners had walked all over the Pilbara seeking fame and fortune, while it sat at their fingertips and right below their feet. Quinton went into lateral thinking mode, and communicated the results to Novo's shareholders. [36] Here is what he said.

> Recent trench sampling revealed a potentially important behavior of oxidized reef material. While collecting samples, the matrix of many reefs was seen to break away from cobbles and boulders as rock was moved during sampling (see Figure 1 below). The Company believes that it may be possible to significantly upgrade material for processing by first removing largely barren cobbles and boulders. If so, this could mean the tonnage of material that will require grinding may be reduced, a potential cost saver. To test this possibility, further test work will include a component of "scrubbing" whereby material will be tumbled to cleanse cobbles and boulders of gold-bearing matrix. A new bulk sample (250 kgs) has been collected from representative reef material on Golden Crown Hill and is being submitted for bench scale scrubbing, gravity and floatation testing. Results from this new test work are expected back in the first quarter of 2015.

(Figure 1: Left, 1m high gold-bearing conglomerate horizon exposed in a trench. Hard siliceous boulders and cobbles occur in a dark brown sandy matrix. Gold occurs in matrix material. Right, loosened conglomerate material collected during trench sampling. Water bottle is approximately 20 cm tall. Note that most of the siliceous boulders are free of matrix material. Screening these out may potentially help upgrade matrix material and help reduce costs of grinding in a mill.)

When faced with the matter of how to mine the conglomerate, Quinton looked beyond conventional drilling and blasting from the start. It would cost too much money to move what was mostly waste rock. Processing it would be inefficient and expensive. He needed a better solution.

He was attacking a problem that had been known about for a hundred and twenty years. Everyone knew there was gold in the conglomerates, and that the conglomerates came to surface in a number of areas. Any prospector could go to a gold-bearing portion of conglomerate and start pulling out gold. In the early days, many did.

But they couldn't extract much gold by hand, and even after

bulldozers and dump trucks arrived after World War II, the gold was too low a grade to be economic. That is, if you did it the same way every other mining engineer or miner would.

So Quinton devised his own solution. It was both simple and elegant, and it worked. He wrote about his theory and how it would work in a press release dated February 9, 2015.[37] He also talked about a deep drill hole. Here are two illustrations and an extract from the text.

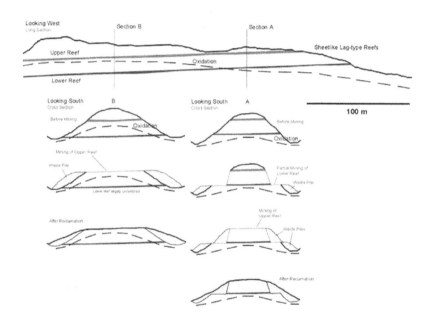

(Figure 1: Conceptual mining plan at Beatons Creek. Overburden above the lower reef, blue, will be pushed aside into adjacent low areas and this reef will be selectively mined. Overburden above the upper reef, red, will then be pushed off to the sides, and this reef will then be selectively mined. This technique reduces the need to haul waste material and could be very cost effective.)

Quinton envisioned stripping off the waste rock and mining only the gold-bearing ore. This was a low-cost mining method, and easy to gain a permit for.

Using simple equipment such as an excavator with a flat

edged bucket, gold bearing conglomerates can be selectively mined thus reducing dilution and, thus, helping maintain higher grades. Costs of this type of mining are anticipated to be low. Figure 2 (below) is a schematic illustration showing this selective mining technique.

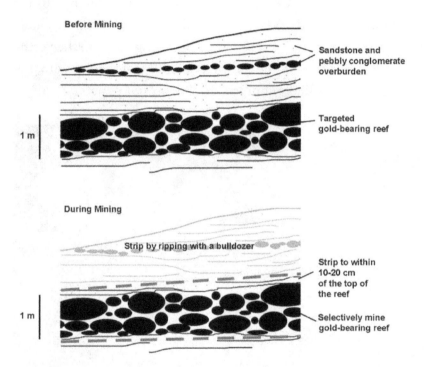

(Figure 2: Schematic illustration showing selective mining concept at Beatons Creek. Using a bulldozer, soft sandstone and conglomerate overburden can be ripped and pushed away exposing the underlying gold-bearing conglomerate horizon. The targeted conglomerate can then be selectively mined utilizing an excavator fitted with a flat edged bucket.

To succeed at Beatons Creek, Quinton needed more land and more gold. After hearing tales of Millennium's desperate financial condition, Quinton went to them with an offer. He would pay them $3.8 million Australian for the remaining 30 percent of the Beatons Creek project. While Millennium owned that 30 percent, Quinton was trapped. If he was too successful developing ounces in the ground, it gave them a life ring. They were too stupid to consider doing any sort of deal with Novo, as they were still convinced Novo would fail. By buying the remaining 30 percent, Novo could take Beatons Creek out of the equation.

Quinton announced Novo's offer late in March, 2015.[38] Millennium jumped on it. They couldn't wait to get their hands on the money. Six days[39] after agreeing to purchase the remaining portion of the property, Novo sat down with the management of Millennium to sign the transfer papers.

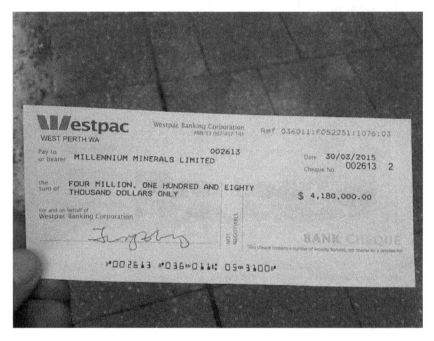

The final payment for Beatons Creek

The meeting concluded with Novo handing over $3.8 million Aussie, plus 10 percent stamp duty. The Millennium lawyer grabbed the check, quickly checked it over and then literally ran out the door to go to the bank. They were in that bad a shape. A few minutes later the bank called to ask if they should honor the check. They weren't used to issuing a corporate check and having it presented in person two hours later. They wanted to make sure the right person had the check.

That was a major coup on Quinton's part. In any mining deal, the health of your partner is important. People get stupid when they have their backs against the wall.

Quinton was still in the position of having enough gold to mine but no mill. Financing and building a mill for the small amount of gold Novo had defined didn't make sense.

But the price of gold kept going down. Eventually Millennium would tip into bankruptcy and Novo could take over its mill.

In May of 2015 Novo moved from the C-Exchange in Canada to the TSX Venture Exchange.[40] That would give the shares more liquidity because in general, small investors are hesitant to trade C-Exchange stocks, believing that TSX Venture shares are safer. I'm not sure how true that is, but the liquidity issue is one that hampers trading in all the companies using the C-Exchange.

About this time, Quinton went to Toronto on business. Customarily he would meet with Eric Sprott, another big Novo shareholder. Quinton wanted to keep him abreast of things.

It was an important meeting, seen in hindsight. It would put over $1.5 billion in Eric Sprott's pocket. It would make a multi-millionaire of Tony Makuch, president and CEO of Kirkland Lake Gold, along with Greg Gibson, Eric Sprott's man Friday. Alas, jealousy would ensure Quinton was stiffed for his contribution.

The highest prices for gold shares were seen in mid-2011, while the top in gold was later, in September. Prices for the metals and resource shares had been tumbling since then, after a twelve-year advance. Many stocks were down by 90 percent and more. The very bottom was at the end of 2015, with the Sprott funds down as much as 93 percent.

A meeting had been held earlier in 2015 to determine just whose head should roll. The verdict was unanimous: Eric Sprott was dumped as the lead portfolio manager on the Sprott equity and hedge funds. This was after being replaced as CEO of Sprott Asset Management in 2014.

While the press releases mumbled about how Eric was in his seventies and wanted to spend more time with his family, and while it soundly praised him for his contribution to mining around the world, the fact remains he was fired. Ultimately, in April of 2017 he was removed as chairman of Sprott Inc. [41]

Eric's old office was on the twenty-eighth floor of the RBC Building in Toronto. Lest he forget his new position, he was demoted two floors to new, smaller quarters on the twenty-sixth floor, also losing much of his view of the city.

When Quinton saw him that day in May of 2015, Eric Sprott was downcast and disheveled, to say the least of it. Normally well dressed and natty, his changed appearance matched his demeanor. He was dejected at being removed but it was impossible to dim his belief in the ultimate fate of gold and gold resources. But he wasn't interested in hearing about Novo or the progress Quinton had made in Western Australia. After all, why should he be? He had little to do with Sprott now.

He hadn't lost his optimism. He still had hundreds of millions of dollars of his own. Eric asked if Quinton would look at properties for him now and again. Of course Quinton agreed. It was his forte.

That meeting would directly lead to the purchase of an Australian mine with the highest grade gold in the world and the lowest cost of production.

For much of 2015 Novo kept advancing the Beatons Creek deposit. They completed the preparation work in anticipation of building a mill, but that was never the real plan. It would be the dumbest thing in the world to spend millions of dollars in planning and permissions to build a $100 million mill that was a stone's throw away from another $100 million mill, owned and operated by a company that had never made a dime mining gold.

Gold continued lower. Sooner or later Millennium would fold.

CHAPTER 14
QUINTON DISCOVERS A RICH GOLD MINE

MUCH OF WHAT OCCURS IN LIFE can be put down to pure chance. Stuff happens because other stuff happened. The meeting with Eric Sprott in March of 2015 would have a long-lasting and far-reaching effect that couldn't have been predicted or even recognized at the time. It just occurred, but the reverberations would last for years.

No one in the mining industry is neutral about Quinton Hennigh and his talent. You love him or you hate him. I have never found anyone in between. If you are secure in your own abilities and judgment, and spend five minutes discussing anything about mining or geology with him, you shake your head in wonder because you just learned something simple that you should have known but didn't. And he explained it so clearly that you got it right away, and realized: I could have done that. But you didn't. So you go back for more because he is so interesting to listen to and knows so much. And he wants to share all that he knows with you.

On the other hand, those who have made it in life and the resource industry through a hearty combination of bluster and bullshit hate him. They wish they were that smart or knew that much, and deep in their black hearts realize they come up short. They can't wait to badmouth him or try to cut him off at the knees, knowing they are no comparison. It's amazing how many bullshit artists populate the resource industry. There are almost as many conmen in it as there are in the newsletter business. Those guys hate Quinton because he makes things seem so simple. So simple they just don't get it.

Eric Sprott called Quinton in early 2016. Eric had been

following the Fosterville gold mine in Victoria, Australia. A company called Newmarket Gold owned Fosterville. Sprott had been accumulating Newmarket shares in anticipation of getting it to merge with Kirkland Lake Gold. Eric had a major position in Newmarket and also owned a lot of Kirkland. He felt that Fosterville had potential for a lot more gold. He wanted Quinton to visit the project and tell him if he should buy more.

Quinton arrived at Fosterville on May 12, 2016. Doug Forster, president of Newmarket Gold, greeted him and took him on a tour. Newmarket had been processing the near-surface gold and at the time were milling 6.91 grams of it per ton of rock. What Quinton saw convinced him that they had a lot more potential than that.

The strange thing is that Fosterville had been showing strong sniffs of high-grade gold for a couple of years. That was what convinced Eric it was a rather large deposit (or BFD, in mining parlance). Another Canadian junior, Crocodile Gold, had owned Fosterville until it merged with Newmarket in May of 2015. [42]

In March of 2014 Crocodile reported assay results from the Lower Phoenix structure,[43] later to be named the Swan zone. It showed exceptional grade and thickness, including 122 grams of gold per ton over 6.8 meters and 24 grams per ton over 33.8 meters. Those are home run holes. But Crocodile didn't understand the nature of the system.

In February of 2015 more assay results came out, including 123 grams per ton over 9.4 meters, 51 grams over 6.0 meters, and 294 grams over 1.8 meters. Crocodile may not have understood the nature of the system and what those grades meant, but it didn't pass Eric Sprott by.

By late 2015 you couldn't even give a high-grade gold mine away. Why would anyone want to own gold? What future was there for it?

High-grade gold in veins from the Fosterville mine

This style of mineralization is called epizonal. These deposits are rare but rich. Just prior to Quinton's visit, Newmarket was drilling deeper holes and coming up with higher grade hits than the six or so grams per ton they had been mining. The gold content was spectacular.

This kind of gold is easy and cheap to recover via gravity. When Quinton saw it on the shaker table he was quite impressed.

He sent his report to Eric. He said that there was a lot of high-grade gold potential at Fosterville. It would require deep drilling but the company was well worth buying. Eric continued to add to the shares in Newmarket that he already owned.

Core speckled with tiny gold bits

They drilled in June. It was deep, almost a kilometer below surface. In July, RBC Bank contacted Quinton and asked him to prepare a due diligence report in anticipation of Kirkland Lake merging with Newmarket.

Quinton flew down in August to do the report and by chance ran into Tony Makuch, the president of Kirkland Lake Gold, who had just arrived from Toronto. They shared a taxi to the project. By this time the assays had come back from the deep drilling in June. When they looked at the core it was obvious they had not hit the high-grade zone. The core showed the quartz stockworks, similar to what Quinton had observed during his May visit.

Gold table, for concentrating gold from the Swan zone

Quinton was highly encouraged and told Doug Forster to drill slightly deeper. Tony Makuch told Quinton that he wasn't about to pay $1 billion for Newmarket, and that as far as he was concerned, they had nothing. It was a piece of dog shit.

When Quinton sent his due diligence report to Eric Sprott he told Eric to not worry about the price; the gold was there. Meanwhile, Tony Makuch was pissing on both the project and the company. Eric was as nervous as a long-tailed cat in a room full of old ladies in rocking chairs, but went ahead due to Quinton's encouragement.

In October of 2016 Kirkland Lake made its offer to merge with Newmarket Gold. The offer was valued at about $1 billion. In Toronto no one thought much of the deal and it only went through at Eric Sprott's insistence. The merger closed on December 1, 2016. [44]

Quinton Hennigh did not discover high-grade gold at Fosterville. Crocodile Gold drilled and found it at least two years earlier. Quinton did pinpoint the Swan zone. That deposit would make the fortune of Kirkland Lake Gold. Quinton's secondary contribution was in understanding the nature of the system and how to advance it. His primary contribution was to hold Eric Sprott's hand and tell him it was a project made for Kirkland Lake Gold, in spite of the opposition from Kirkland Lake's management.

On January 17, 2017, just six weeks after the merger took place, Kirkland Lake announced some of the highest-grade assays ever recorded in Australia, and certainly the highest ever at the Fosterville mine. [45] The best was 1,429 grams of gold per ton over 15.1 meters, including an incredible 21,490 grams over 0.6 meters, in hole UDH1817.

It was exactly as Quinton Hennigh had predicted in August. I can't tell you how many people that pissed off.

The Lower Phoenix Footwall was promptly renamed the Swan zone. The head grade for Fosterville went from 6.9 grams of gold per ton to 49.6. It became the richest and lowest-cost gold mine in the world.

At the time of the merger, Kirkland Lake Gold was valued by the market at about $2.4 billion for all its projects. It now carries a value of $16.64 billion. The shares went up over 1,000 percent and made Eric Sprott another one-and-a-half billion dollars.

Tony Makuch and his sidekick Greg Gibson both made millions on their now ultra-valuable options. Neither would forget that Quinton Hennigh made that money for them, and made it look easy.

The relationship between Kirkland Lake and Novo would become more and more interesting, with Kirkland helping Novo at important points and nearly destroying Novo at other times. I seriously doubt that the board or Eric Sprott ever knew about the games being played behind the scenes by Gibson and Makuch.

CHAPTER 15
SNIFFS OF GOLD AT KARRATHA

EVER SINCE QUINTON CAME UP WITH HIS THEORY of how the gold got into the Witwatersrand, he saw the Pilbara region as potentially hosting billions of ounces of gold. He always wanted to go big rather than have just a deposit or two. As he advanced Beatons Creek and Marble Bar he continued to seek out other properties that would work with what he already controlled. He had no intention of breeding and training a one-trick pony.

In August of 2015 Novo announced its intention to purchase what it called the Blue Spec and Gold Spec[46] deposits from Northwest Resources, an Australian company. This would add about 215,000 ounces of high-grade resources to its portfolio. Novo paid with cash and shares. By 2016 the company had begun to devote resources to investigating just what the potential might be at Blue Spec.[47] The two deposits were located about twenty kilometers due east of Beatons Creek, so Novo personnel could easily support geological surveys and work at both projects.

Most of the groundwork necessary to develop a mine and build a mill is boring — way too boring for most investors, who want something interesting, and immediate action. Punters treat a purchase of a junior mining company's shares as they would any other lottery ticket. If the draw will be on Saturday night, they want something happening on Friday and Saturday. But building a mine and a mill can take a decade or more.

Novo advanced Blue Spec and another newly acquired property named Talga Talga[48] during 2016, while moving Beatons Creek towards trial mining. In the background, events began to take place that would have giant implications for Novo and for Quinton.

In the late summer of 2016 Quinton began hearing about the discovery of patches of gold nuggets around the town of Karratha in Western Australia. That's at the west end of the Pilbara Basin.

The best geologists have a snout like a beagle pup. They stick their noses into everything in the hopes of finding something interesting to gnaw on.

Sometime in September, Bill and Kerry Edwards contacted Quinton and reported that someone on a crew digging gravel for some Rio Tinto rail work saw gold nuggets in the gravel. The workers promptly spent $10,000 on a metal detector and returned to the gravel pit on the weekend. That weekend was worth a quick three hundred ounces of gold. I'm certain it was reported to the government and all taxes were paid on it.

Bill and his wife Kerry lived across from the Novo camp at Nullagine. Bill did a lot of small-scale mining and kept his ear to the ground on Quinton's behalf.

Over the past twenty years I have met a lot of geologists. I suppose I have crossed paths with a thousand or more. Most cast no shadow. I'd say that very few of them were much better than shake-and-bake geos who could recite what they were told in class and hadn't had an original thought since their first visit to Lima, where they were trying to figure out where and how to get laid. That's pretty dumb. If you can learn how to fall off a bike you can figure out how to get laid in Lima.

But I have also gotten to meet and talk with the very best in the industry, such as David Lowell, Peter Megaw, Keith Barron, and Quinton. The greats are interested in everything, have a wide range of knowledge about just about anything, and are constantly trying to piece things together.

I've met all these guys and have spent a lot of time with them over the years. All things fascinate them, and each can talk intelligently on dozens of subjects, not just rocks. And they ask questions constantly. They question everything, and perhaps more importantly, they listen. Any opinion might have some value. I've been very lucky to know these guys.

The Pilbara Basin is home to one of the largest and richest banded iron formations in the world. According to Wikipedia, about 59,500 people are employed in iron mining in the Pilbara. Gold mining dates back to 1888 through about 1900, and it would be safe to say that over the years, 100,000 miners of one sort or another have trodden on the Pilbara in search of valuable minerals. It took that rail crew from Rio Tinto, Johnathon Campbell, who I will return to in a moment, and Quinton to put it all together.

The rail crew needed only to notice that the pit they had opened up for fifty truckloads of gravel contained gold, in order to transfer three hundred ounces of it into their pockets over two days. That information told Quinton that he needed to get busy staking, before the whole world realized the potential of the conglomerates. If one pit about two meters deep and perhaps one hundred meters across can generate $500,000 worth of gold in a weekend, how much more gold is still around there?

The point here, and it is an important point, is that the gold discovery at Mallina by Jimmy Withnell took place 129 years before the light switched on in Quinton's brain. Government reports from the early 1900s mentioned gold in the conglomerates and in the eroded gravel. The existence of the gold was established all those years ago, and no one put it together until April of 2017.

On November 16, 2016 a tiny Australian junior named Artemis Resources issued a press release announcing that it had found gold nuggets at surface at its Purdy's Reward project.

Here is an extract, including one of the Figures referred to.

ASX Announcement
16 November 2016
Purdy's Reward Gold Discovery – Karratha

Highlights

- ✓ Visible gold and nuggets exposed in mafic rocks, at Purdy's Reward Project 35 km south–south east of Karratha.
- ✓ Surface gold identified over a potential 800 metre strike.
- ✓ Gearing up of exploration activities

Artemis Resources Limited (ASX: ARV) is pleased to announce that recent exploration activities have confirmed the presence of primary gold mineralization (Figures 1 and 2), with significant free gold, in mafic rocks 35 km SSE of Karratha ("Purdy's Reward Project").

The primary gold mineralization was recently discovered by prospectors in the belief that, because the gold was flat and rounded, it was elluvial in nature. The visible gold actually sits within weathered mafic rock and requires significant hand pick, crow bar and sledge hammer work to liberate. Free gold has now been found over a strike length of 800 metres with widths up to 100 metres within the project area (Figures 3, 4 and 5).

The geology of the project is characterized by Archean felsic and mafic rocks. The only previous exploration work in the area was back in 1971 by Westfield Minerals NL and this exploration programme focused only on base metals exploration with 6 percussion drillholes on the western tenement boundary. These holes were assayed for nickel with assays returning up to 1260 ppm Ni in drillhole 69-SP-07 (Table 1, Figure 5) associated with a chloritized mafic basalt.

Figure 1. Artemis, Purdy's Reward Project - Karratha. Gold nuggets, flat and rounded and up to 13 grams from surface detecting.

All gold mineralization observed and found to date in the

West Pilbara has been associated with quartz reefs. This new style of gold mineralization within mafic hosted rocks increases the potential size of mineralized horizons.

> Ed Mead, Artemis's CEO, commented: "The initial work from Purdy's Reward increases our confidence in this under explored gold region of Western Australia. Off the back of recent results from Silica Hills, the best indication of gold is visual gold at surface, and we certainly seem to be getting that."

Most geologists will confirm that visible gold at surface is indeed a strong indication of the presence of gold. Pedants may wish to discuss the precise meaning of the expression "certainly seem".

For the next couple of years Ed Mead would claim that it was he who discovered gold in conglomerates around Karratha. He also claimed that the Witwatersrand basin contained "watermelon seed gold identical to that found in the Karratha region." Alas, neither claim was true.

In Geological Survey #33, published in 1909 by the government printer in Perth, a geologist named A. Gibb Maitland stated that "nearly all the alluvial gold was of the uniform shape and size of small melon seeds." He was talking about gold from the Egina region, but the gold from Karratha proved to be identical. Another Geological Survey published in 1906 suggested that the source of the gold in the Pilbara was the conglomerates.

Quinton Hennigh realized how significant that was. The distance from Egina to the projects he would pick up at Karratha was about 115 kilometers. And the gold from the two areas was virtually indistinguishable. The gold from the Witwatersrand was in tiny bits, and there was no watermelon seed gold at all.

Credit for any gold rush belongs rightly to the person who made the discovery that people noticed. It was known as early as 1842 that there was gold in California, but it was John Marshall's discovery at Sutter's Mill in January of 1848 that brought about the incredible mass movement of prospectors to the new state of California.

Likewise, in the Yukon, it was George Carmack and Skookum Jim who started the snowball rolling down the mountain after they found gold on August 16, 1896 at Rabbit Creek, later renamed Bonanza Creek, more appropriately. Their discovery began the short-lived Klondike gold stampede of 1897–98. But the natives had known about the gold for centuries. They had no use for it, and much preferred the copper that could also be found there.

The person who actually began the Karratha gold rush was a cattle rancher named Johnathon Campbell. In August of 2016 he and Bruce Woods, a Kiwi helicopter pilot, took a Robinson R-44 chopper up near Karratha to count wayward cattle. Following a creek system downstream, they passed over a camper hidden under some trees. They supposed that it was someone either on the run from the law or prospecting for gold. In Australia that is often the same. As they continued down the creek they saw a number of systematic holes poked in a gully leading from the dried-up creek bed.

Campbell lives near Port Hedland. He placed a notice on Facebook, looking for a gold prospector to work with. A man named Brad Smith answered his ad. They purchased metal detectors and went back to the area that Campbell and Woods had flown over.

Campbell and Smith drove to what is now called Purdy's Reward, pulled out their metal detectors and went to work. They found gold nuggets everywhere they looked. Brad Smith soon caught gold fever and commented that he had never seen anything like it in his life. That comment brings back memories of watching the movie, *The Treasure of the Sierra Madre*,[49] and what happens to people when greed for gold overtakes good sense.

On that first visit, the two intrepid prospectors spent only half a day in the field but took home about $5,000 worth of gold nuggets. Smith was hooked, especially when Campbell promised to split everything they found. On subsequent weekends spent prospecting on their claims they would take home between $5,000 and $20,000 every time.

Australia has an interesting mining right, designed for prospectors and small miners, called a Special Prospecting Licence.[50] It allows prospectors to mine for gold over a mining lease held by someone else, on a limited basis. Johnathon Campbell applied for and was granted SPLs on both Purdy's Reward and Comet Well. Brad Smith applied for a small block further down the creek but soon dropped it, saying it was no good and a waste of money. Campbell picked it up but was mocked as a fool by Smith.

Artemis Resources held the lease on Purdy's Reward, and a man named Peter Gianni owned the lease on Comet Well. The SPL allows mining on a small scale if the primary leaseholder does not object.

Not long after this, on another chopper flight seeking lost cattle, Campbell and his pilot flew over some people operating a small bulldozer to scrape the ground. They were then following the bulldozer with a metal detector and collecting the gold nuggets they found. Campbell and his pilot landed and talked to them. It was Ed Mead and one of his people from Artemis.

Artemis had gotten into hot water with the Department of

Mines and the local native group for disturbing the ground without any permit or agreement with the natives. Claiming they were rehabilitating the ground, the pair was instead inflicting even more damage. They freely agreed they were selling the nuggets to the Perth Mint.

I have looked at Artemis' financial statements. Nothing is obvious about any revenue from illegal gold sales. If it had been following the rules, it would not have been permitted to sell the gold in any case until it had obtained a mining license. Which it didn't have.

Campbell asked Mead who the registered holder was of the "Armada" lease. Mead volunteered that Artemis held the property. Campbell said that he had been trying to contact the company, but no one ever picked up the phone or returned calls. Ed Mead said the company had no money, and that for all purposes, he was the receptionist.

In a neighborly spirit, Campbell told Mead that he had applied for an SPL over the part of the Armada lease which came to be called Purdy's Reward. Ed said he didn't care; he had no interest in the ground other than Artemis controlling it from a legal point of view.

Campbell persisted, and told Mead that it was far better ground than they were working. Mead should at least look at it. Remember, he was claiming to be doing rehabilitation work, and that that was why the bulldozer was there. Artemis had never done a thing with Purdy's Reward, so claiming to be cleaning up past work wasn't just weak, it was impossible. He was using the bulldozer to find gold.

Mead said he didn't care if Campbell worked the SPL over Purdy's Reward, but changed his mind in a hurry once he ran a metal detector over the project and took out a lot of nuggets. He immediately chased down Brad Smith, who was camping nearby, and told him to get off the ground and not to return.

As for Johnathon Campbell, to this day he thinks about Ed Mead as a street he has driven down in Perth. It's marked "ONE WAY." Mead would have never known there was gold at Purdy's Reward, but for Campbell.

Smith and Campbell, through their SPL, still had the right to prospect for and find nuggets at the Comet Well project. One day in late 2016 or early 2017 they talked to a geologist named Rob Jewson about trying to buy the lease on Comet Well from Peter Gianni. Brad Smith was hard to get along with, and people who dealt with him tended not to want to do it twice. Gianni agreed to take $50,000 for the lease but wouldn't deal with Smith.

Then one day Smith called Campbell to say he had found someone who would put up $100,000 for the lease on Comet Well. The three of them would own equal shares of the project. The new investor was named Darren White but everyone involved always referred to him as YT.

Campbell was in agreement and quite willing to sell an interest in the pending lease to White, as was Smith. But when the terms were first discussed in January, Brad Smith refused to put his half-interest in the SPLs on Comet Well into the deal. He told Campbell not to consider adding the SPL to the deal. He could take out thousands of dollars' worth of gold every weekend, so why should he share it with someone new?

But after the sale was concluded in April of 2017, with Darren White now owning half of the Comet Well lease, he demanded Campbell throw his interest in the SPLs into the deal. By that time Smith had flip-flopped, rewarding the guy who had put tens of thousands of dollars into his pocket by joining with White and changing the terms of the deal after it was signed, sealed, and delivered. Campbell reluctantly agreed, but was starting to figure out which way the wind was blowing.

Quinton made his first visit to the Comet Well property, arriving on April 6. What he saw convinced him to get in touch with the leaseholders. YT didn't want Campbell talking to Quinton. Quinton sat down with YT and made him an offer. Fifteen minutes after leaving YT's office, Quinton's phone started ringing. It was Mark Creasy. Quinton didn't answer any of the calls.

Obviously, and as Quinton later confirmed, YT had taken Quinton's offer straight to Mark. Mark made YT an interesting counteroffer. He would buy all of Comet Well for $5 million by paying $50,000 for each percentage point. He would hand over $50,000 to the trio for Comet Well, gaining the right to test it and collect gold without paying one cent more. If it proved valuable, he would buy the additional 99 percent. If not, he would walk.

It was not only a predatory offer. Mark was a Novo insider due to his share position, and there is the interesting question of how he could become involved in negotiations between Novo and a third party, ethically or legally.

YT eventually accepted a higher bid from Novo, but YT talking to Mark Creasy greatly increased the cost to Novo. Dealing with the three owners of Comet Well was difficult because none of them could agree with the other two on how to proceed. Eventually Quinton got Johnathon Campbell to accept an offer for his one-third share. That forced Smith and YT to come to some kind of reasonable terms with Novo. Quinton does not reminisce fondly about dealing with three people who had entirely different views of what would constitute an acceptable deal.

What was interesting to me was that Comet Well was a free ride for Brad Smith, who took out tens of thousands of dollars in gold within just a few weeks after being pulled into the deal by

Johnathon Campbell. YT bought into the project in February and had an agreement with a well-run and well-financed junior exploration company two months later.

The back story of the gold is far more interesting than the above somewhat confusing press release issued by Artemis in November, when Ed Mead began claiming to be the fellow who began the Karratha gold rush. First we must identify some of the more important players.

David Lenigas became executive chairman of Artemis Resources on November 3, 2016. His timing was impeccable. Artemis was a junior with a tiny market cap, no money, no focus, and a zillion shares outstanding. His appointment came with a $5,000 monthly salary and, as an incentive, twenty-five million shares of the company.

All he had to do to make a fortune was to increase the value of the company to an absurd level, sell his shares and ride off into the sunset, leaving the rubble of Artemis in his wake. That is exactly what he would do, over time.

Lenigas was your typical overweight, pompous, beady-eyed blowhard, living in Monaco off the rewards from the other 160 tiny juniors he had churned and burned. In physical appearance and in attitude he resembles Goldfinger, the antagonist of the 1959 novel and the 1964 movie of the same name.

He fancied himself as the reincarnation of Tiny Rowland [51] and indeed there were similarities, including a willingness to stiff shareholders regularly. In mining I have found that some management types mine rocks, and some mine shareholders. Lenigas mined shareholders again and again.

When he became chairman, Artemis was a company without a clue. It claimed to be a copper company and a cobalt company and a nickel company. It intended to mine zinc, lead, platinum, diamonds, obviously gold and silver, and even had a graphite project in Greenland that fit its portfolio perfectly. In other words, management had no idea what they were doing. They would soon spend $3.5 million on a rusted wreck of a processing plant that would forever be about to go into production.

Artemis was perfect for the Lenigas magic. He even had a mysterious investor in the background, Mick Shemesian,[52] dubbed "Many Names" Shemesian by the Australian press due to his refusal to respond using the same spelling of his name to the many lawsuits in which he was involved. Shemesian was the largest shareholder in Artemis.

Allan Ronk, who had worked with Quinton at Beatons Creek, was on an assignment for Artemis in the Karratha region. When he learned about the gold the rail crew had found and saw a February 20, 2017 press release[53] from Artemis, he became the first to connect the dots. Below is a lengthy extract from that press release, plus one of its six Figures.

While Ed Mead may have believed all of the gold in the Pilbara was in lode gold veins, Quinton was well aware of the government reports going back over a century. He was convinced that the key to the gold jewelry box was in the conglomerates. If that was true in the Eastern Pilbara, as it surely was at Beatons Creek, why couldn't it be true in the Western Pilbara, 360 kilometers to the northwest?

ASX / Media Announcement
20 February 2017
New style of gold mineralisation for the West Pilbara discovered at Purdy's Reward gold project – Karratha, Western Australia.

- Archean sediment hosted gold mineralisation identified.

- Typical sediment style deposits of this type are Witswatersrand, Nullagine and Marble Bar which are conglomerate hosted.
- Assays confirm gold mineralisation is typical of this style with fine to coarse gold.
- Geochemical assays from orientation traverse identifies gold anomalies coincident with high levels of prospector activity.
- Free gold continues to be found at surface at Purdy's Reward Gold Project by prospectors.
- Programme of Work for drilling and costeans has been approved by the Government.
- Heritage survey requested and expected in March.

David Lenigas, Artemis's Chairman, commented;

"The results from the new Ionic Leach geochemistry techniques have now been received and they confirm a completely new style of gold mineralisation for this area of the West Pilbara akin to the Witswatersrand style Archean sedimentary hosted deposits. These deposits are similar to those being discovered in the East Pilbara, around Marble Bar and Nullagine, by TSX listed company Novo Corp. Novo's work has identified predominantly fine gold mineralisation, whereas Purdy's Reward not only has fine gold but coarse gold in the form of nuggets. This makes the Purdy's Reward discovery significant for the area and highly prospective. This is the first time that this type of geology and gold mineralisation has been confirmed as Archean sedimentary and associated with conglomerates and fine grained sediments with mafic appearance."

Artemis Resources Limited ("Artemis" or "the Company") (ASX: ARV) is pleased to report the identification of a new gold mineralisation style for the Karratha area at Purdy's Reward, West Pilbara.

Results (Figure 2, 3 and 4) from a geochemical orientation sampling traverse at the Purdy's Reward Gold project, and additional mapping support a model of Archean sedimentary (conglomerate) hosted gold.

The gold occurrence at Purdy's is considered analogous to the conglomerate hosted mineralisation outlined by Novo Resources in their Beaton's Creek Project near Nullagine, but Purdy's Reward is significantly older in age. The style of mineralisation is referred to the Witswatersrand style, after the Witswatersrand gold province in South Africa that has significant gold in Archean sedimentary conglomerates.

The traverse straddles the unconformity between the older Archean basement in the north and the overlying the Mt Roe Basalt to the south.

The samples were collected along a 1 km long traverse; samples for analysis using the ultra-sensitive ALS Global Ionic Leach™ technique were collected at 25 metre intervals. For comparison purposes at every 100 metre sample point an additional sample was collected and analysed using a conventional digest (Supertrace).

Within the central area of the traverse numerous prospector metal detecting pits were present. The area of the detecting pits corresponds to where significant anomalism in both the Ionic and Supertrace results occur. The results are expressed as "Response Ratios" for both techniques. This is to enhance the response to background signal of the data and to perform basic levelling on the results; this allows direct comparison of the results from the different methods. As such these values do not have a unit type.

Figure 1: New Conglomerate Package identified at Purdy's Reward:

[Figure 2 shows Purdy's Reward orientation geochemical sampling traverse.]

The Ionic data (Figure 3) shows a stronger anomaly to background response compared to the Supertrace data. Neither technique (Figure 4) shows significant correlation between the strong gold responses and the typical pathfinder elements: arsenic, silver, bismuth, molybdenum, antimony and tungsten.

This lack of pathfinder elements (Figure 4) indicates the gold is not derived from a shear system or from supergene alteration of a shear system. The data suggests that the Purdy's Reward gold is from an alluvial source, albeit Archean in age. The lesser gold responses on the right hand side of Figure 4, which are at the northern end of the traverse (Figure 2) do show coincident gold, silver, arsenic responses indicating there is a secondary zone of mineralisation in the area which is shear system related.

Allan Ronk told Quinton to take a close look at that press release. Allan had of course identified the mineralization as being identical to that of Beatons Creek and Marble Bar, and had provided most of the technical information in the press release.

At this time, literally no one had paid serious attention to the conglomerates for over a hundred years. In the Beatons Creek area, miners in the nineteenth century actually dug caves into the conglomerates to mine for gold, but no one had investigated and then said, "Maybe all the conglomerates hold gold" — except for Quinton Hennigh, and, generations before him, a few government geologists who were largely ignored. Everyone believed that virtually all the gold was from a lode source.

I should mention one confusing detail here. Frankly I don't know why the conglomerates are called "reefs" in the Witwatersrand but "conglomerate layers" in the Pilbara. I like reefs better.

Quinton looked over the press release from Artemis, with its pretty photo of a conglomerate outcrop, and took action at once. Novo's exploration manager at the time was Luke Meter. Quinton told him to pack his bags and go to Karratha and peg every claim around the area where the conglomerates showed at surface.

By April 2017 Novo was down to $1.5 million. Ronan Sabo-Walsh, its financial guy, was sending $100,000 to Perth each week to cover expenses incurred in filing the claims. Quinton couldn't talk about the new properties as it was a work in progress; Novo was still busy staking as much ground as possible. But it would become a great deal for investors within a very short time.

A private placement was being done at sixty-six cents a share, with a full warrant. Initially Novo planned to raise $8 million, but the terms were so attractive that the size of the placement was first increased to $12 million Canadian, and then to $15 million by the time it was closed in April 2017. Those wise enough to participate were in for the ride of their lives by the time the four-month holding period was over.

Novo was still busy on the eastern side of the Pilbara Basin. It had begun a 30,000-ton bulk sampling program in mid-2016 and was busy drilling in the Blue Spec. The results from the bulk sample showed the grade at Beatons Creek to be much higher than the numbers obtained from the drilling and surface samples.

Also in April, Novo announced the deal on Comet Well. All this time, behind the curtain, Novo's geologists were staking land non-stop. On the strength of the Comet Well agreement, Quinton was able to reach agreement with Artemis on a package of just over 1,500 square kilometers, including Purdy's Reward.

Lenigas actually called Quinton to see if Novo was interested in doing a similar deal on Purdy's. Lenigas didn't understand it at the time and perhaps still doesn't, but to Novo, the primary attraction of Purdy's wasn't the quality or size of the project. Artemis had something far more valuable: a drill permit for Purdy's. Getting the same for Comet Well would probably take a year. Dealing with the bureaucrats in Australia is like swimming through cold molasses. It's wet, nasty, and slow.

Mead and Lenigas still believed the bulk of the gold would be in lode gold veins, so they thought they were restricting Novo by making the agreement valid only for the gold in the conglomerates. But that was all Quinton was interested in.

In the press release announcing the binding letter of agreement on Purdy's Reward, Novo mentioned for the first time the staking of over 6,000 square kilometers of ground; over 2,300 square miles. It was a giant position and was news to the market.

In the press release Quinton explained what Novo sought, saying, *"The basis for staking such a large land package is the recent recognition of gold-bearing conglomerates in a previously unexplored sequence of rocks near the base of the 2.7–2.85 billion year old Fortescue Group, a thick pile of sedimentary and volcanic rocks underlying vast portions of the Pilbara region."*

That was the basis of a giant shift in thinking that most of the tiny juniors in the Pilbara region have still not come to grips with. It was a shift from exploration for lode gold sources to a search for gold from the conglomerate structures. Quinton upended the thinking of the gold community. It would take years for the rest of the herd to realize that the easy gold was in the reefs and the gravels extending to the Indian Ocean.

Immediately after completing the deals with the Comet Well threesome and with Artemis Resources for Purdy's Reward, Quinton made a welcome addition to the Novo management team. He is so talented and smart that he tends to want to do things himself. It's not micromanagement but he does take on more than he should at times, because he can often do a better job than anyone else. That's all well and good, but that kind of management style limits a company to what one person can accomplish, no matter how well.

I had been bugging Quinton for months about a matter he had realized himself but hadn't addressed. As an American running a Canadian-listed mining company in Australia, no one was going to go out of their way to do him any favors.

Western Australia still resembles the Wild West days of the United States and Canada. Everyone knows everyone else's business. To a certain extent it's a closed community, even though Quinton had worked for Newcrest, an Australian mining company, and was highly respected in the country. He was still a furriner.

Quinton needed a first-class manager who was Australian. He announced the hiring of Rob Humphryson as CEO. Rob came with a resume as long as your arm and had experience of a broad range of duties over his twenty-five years working in Australia. Best of all, he was Australian; he understood the culture and knew how to get things done.

At the same time, Quinton promoted Ronan Sabo-Walsh to Chief Financial Officer (CFO). I've known the pair now for several years and both are brilliant. Especially given the COVID-19 lockdowns and stupidity, Novo was now well prepared with a widely experienced management team.

A lot of what Quinton does and announces merely flickers through the minds of investors, seen and soon forgotten. However, he is always playing three-dimensional chess and there is always a method to his madness. Investors and outsiders like to throw rocks at him but his moves are always carefully thought out.

One of the biggest dangers that any junior mining company faces is that of succeeding. As soon as a junior makes an announcement of a big find, everyone wants to steal it, including all of their big partners.

In the early days, when capital is almost always the biggest problem, finding investors with deep pockets seems like a good thing. Having Mark Creasy as a major shareholder was good for a lot of reasons. It also could have been the kiss of death to Novo, and in a moment we will see an example of a major investor trying to chop the company into tiny pieces.

Mark had held big positions in other tiny juniors, and when he could hold a company to ransom he was quite prepared to do so. It was like keeping a pet lion in the back yard. It keeps the neighbors relatively honest, but when it is hungry it will eat your leg without a second thought.

So in all of Quinton's dealings with everyone, he kept in the back of his mind that even the biggest shareholders might turn on the company. There were a couple of times when Mark could have done things to put Novo on the map instantly, but he always put his interests as an individual in front of his interests as a shareholder. I thought that was especially dumb, and I like Mark a lot.

The major mining companies can be dangerous too. Everyone wants someone else to do the work but everyone feels they are smarter than management. Or they see a chance to pick up a prime asset cheap. Big companies, small companies, big investors, small investors — everyone is smarter than management.

If you don't believe me, go to Stockhouse or CEO.CA or HotCopper and see all the fools hiding behind their aliases. They know everything about mining, and if you were to turn any company over to them for a month they could sort out all its problems.

Quinton Hennigh did a better job of keeping his potential enemies at arm's length than anyone I have known in mining. He did the same with his friends, for you never know who might turn.

It's a finely tuned balancing act; you don't want to be offending anyone, but you still have to run the business and make progress. If any single investor gets too big a position, it is only natural that they start thinking they should have the majority of the input as well.

Few management teams make it from discovery to production without being sniped at, and often hit and killed. Everyone wants the bonanza deposits with the least effort and expense.

I made a trip to Japan in June of 2017, to Hokkaido, to visit Irving Resources' new discovery there. Quinton serves as an important advisor to the company. He brought the lovely Heather. Brent Cook joined the tour, as well as Maurice Jackson from Proven and Probable.

While we were in Japan, Quinton showed us some samples of the gold Novo was finding on the western edge of the Pilbara region. It is highly unusual. They are gold nuggets but in a hard rock environment.

Gold nugget from Karratha

A week later, Brent joined Quinton and me on a visit to Karratha, to see that new discovery. The three of us met in Perth for drinks. Quinton showed us samples of gold that he had bought on top of a piece of polished conglomerate.

Gold nuggets Quinton picked up on a previous trip

It was one thing to show us a few gold nuggets in a hard rock matrix while we were in Japan, but Quinton was holding out on us. Once in Australia, we began to get an idea of the amount of gold involved.

There were several prospectors with Special Prospecting Licenses who were collecting gold found with metal detectors. We went out into the field and watched a fellow named Rob Beaton detecting and finding lots of nuggets. Rob was an ordinary prospector with a metal detector who had filed several SPLs and was busy making his fortune. We could legally buy gold from him.

Rob Beaton had lots of gold nuggets for sale

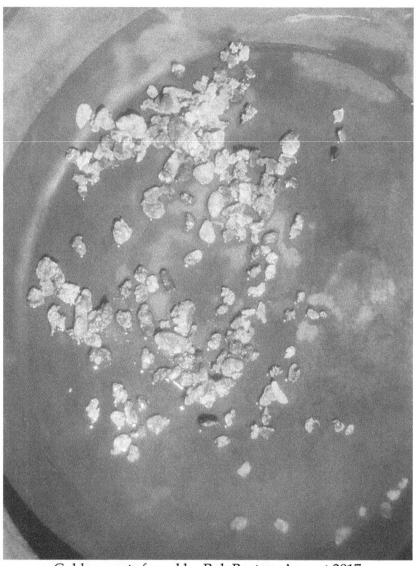

Gold nuggets found by Rob Beaton, August 2017

This was unlike anything I had ever seen. The source of the nuggets was the conglomerate, but unlike Beatons Creek and Marble Bar, the material it was found in was gravel that had been covered in basalt and solidified into hard rock. These were gravel reefs 2.8 billion years old, now metamorphosed into hard rock. Other than the size of the gold, this was very similar to the gold from the Witwatersrand Basin.

I never knew if Rob Beaton was connected in any way to Beatons Creek, some 350 kilometers to the east, but it's a small country and he probably was related.

There were a number of people on the trip. Just after we left the highway we passed the famous gravel pit where the railroad construction crew had made their fortune in one weekend. It wasn't a giant pit, maybe one hundred meters across, but good for three hundred ounces of gold in two days.

Artemis using a small digger to make payroll

We drove out to Comet Well and passed by Ed Mead, who was using a small digger to collect his paycheck for the week. This was the sort of thing that Quinton and Novo wouldn't dream of doing. It was unquestionably illegal but that presented no problem to Ed Mead, Artemis, or David Lenigas.

Novo hadn't progressed to exploration yet. We were visiting Rob and inspecting Novo's ground at Comet Well to see the size of the project. Quinton had his guys stake an incredible total of 7,600 square kilometers, or 2,950 square miles. Gold here at any grade would make this the largest gold find in history.

Rob Beaton detects while Eric Sprott watches

Many of us on the tour bought some gold from Rob. He had a lot of gold.

Quinton and I returned to Karratha and visited the local representative of Minelab metal detectors, reputed to be the best on the market and the most sensitive. At the time of our visit, in the winter of 2017, Minelab's most expensive product cost about $10,000 Australian. There has been so much demand since then, largely because of the incredible stories coming out of Karratha, that the quality of the machines has increased markedly and the price has been chopped in half.

The dealer, located in a dingy hut in a beat-up industrial park, brought out tray after tray of gold. He was so casual about doing so that it was plain he had a lot of gold. We paid a slight premium for what we bought, but it wasn't much. He must have been selling millions of dollars' worth of metal detectors and equipment each year, and taking even more back in purchases of gold nuggets.

He took care of Quinton and me because Novo had brought a lot of welcome attention to the region. He was selling detectors with one hand and filling up the other paw with nuggets, and making a nice margin on both.

He finally brought out a tray loaded with pairs of matching nuggets that he had purchased from a miner with a project near Egina. I would mention to Quinton that the gold from Comet Well that we bought from Rob was identical to that gold from Egina. That implied a giant strike length and a lot of gold.

What I didn't know at the time, and wouldn't learn until much later, was that while Purdy's Reward and Comet Well were both hard rock with embedded nuggets, Egina was an entirely conventional alluvial project. Again, since the Pilbara has either too much water or too little, Egina had been mined off and on for a hundred and twenty years.

The Karratha projects were gold nuggets in conglomerate, while the Egina area was a lot of gold nuggets in gravel. Come back in a few million years and they too would be hard rock.

Gold nuggets from Egina, on display in Karratha

What I also didn't know at the time — because he keeps his cards close to his vest, even from me — was that Quinton had opened discussions with a man named Carl Dorsch, to buy his fully permitted gold project at Egina.

Earlier in this book I mention that years before, I had asked Quinton where the gold at Beatons Creek went when the outcropping conglomerate eroded. He told me that it had been washed into the Indian Ocean.

Well, not all of it had made it all the way to the water's edge. Some of it landed near Egina and was slowly working its way west.

CHAPTER 16
QUINTON FLUBS THE FIRST ASSAYS AT PURDY'S

THE YEAR OF 2017 WAS A LUCKY ONE for Novo and Quinton Hennigh, with a lot of breaks coming their way. Not all of luck is good, however. Quinton flubbed a giant set of assay results and got beat up for years as a result.

When Novo did the placement first announced in March of 2017, it expected to raise $8 million at sixty-six cents a share, with a full warrant at ninety cents. Quinton knew the company's focus was shifting to the Karratha conglomerate gold find but couldn't discuss it while the placement was underway.

Eric Sprott already owned over four million shares and four million warrants, and topped up with another five million of each. Calculated from the total number of shares issued and outstanding, Sprott owned 8.5 percent of Novo. If he were to exercise all his warrants he would own 15.6 percent.

Little did he expect that his position in Novo would soon make him well over $100 million. Remember, he profited by over $1.5 billion at Fosterville, largely because of Quinton's genius in pinpointing the Swan zone and nudging Eric to merge Kirkland Lake with Newmarket Gold. For Eric, it was a great year. For Quinton, not so great.

Quinton signed and sealed the deal with the threesome on Comet Well in April 2017, and the binding letter of agreement with Lenigas on Purdy's Reward in May. In July Novo announced a leap forward: the Sumitomo Corporation had agreed to work with Novo to put Beatons Creek into production.

That agreement didn't look like much because it didn't bind Sumitomo to anything, but anyone who understands how the Japanese do business would recognize just how important it really was. The Japanese look years and years ahead. Sumitomo was in effect becoming engaged with Novo and expected a long and profitable relationship.

Even when Eric Sprott makes a major investment in a company, he rarely has time to hold hands with management. He was now sitting on nearly twenty million Novo shares, which even for Eric was a pretty good size position. He decided he needed someone to represent his interests on the board of directors. In July of 2017 Novo announced it was adding Eric's man Friday, Greg Gibson, to the board.

Naturally Gibson had to be taken care of, so he was awarded half a million options at $1.57. He would sell his shares in early 2019 and clear something like $750,000 to $1 million for less than two years' work. Not all of his work was to Novo's advantage, either. In fact some of it came close to destroying the company.

There was also a side deal between Newmont and Eric Sprott. While it didn't affect Novo at all, it was interesting in that it shows how big finance works.

To recap: Eric had participated in a placement, buying Novo shares for sixty-six cents apiece. The placement called for a full warrant at ninety cents. Quinton knew of things in the works at Karratha, but no information was released to the public or to potential investors. It was a deal of a lifetime for Eric, as those shares would increase in price by 1,400 percent in six months.

But after Novo announced that it had picked up a lot of ground around Karratha, Eric thought it might be nice to own more shares. He asked Quinton if there were any blocks for sale.

Quinton contacted Newmont in Denver. Newmont could reduce the cost of its initial investment in Novo to zero by selling some of its shares. Newmont contacted Eric and agreed to sell him eleven million shares of Novo at $1.56.

Meanwhile, in Perth, the Newmont guys there went out to

Purdy's Reward, saw the trench where the samples had been taken, and became most enthusiastic about Novo's potential. Somehow, the Newmont guys in Perth learned that the Newmont guys in Denver were thinking about selling Novo shares. Perth told Denver that this was inadvisable, as Novo looked to be on to something big in WA.

Eventually Eric Sprott realized that Newmont hadn't invoiced him or made the agreed transfer of shares to him. He asked its Denver office where his shares were. Newmont replied that after careful consideration, they thought they might want to hold on to those shares. Since the first discussion about the sale of the shares, the price had gone up a lot.

Eric had his attorney draft a nastygram to Newmont reminding them that they had agreed in writing to the sale, and to the price. Did they really want a giant nasty public lawsuit on behalf of a billionaire suggesting that Newmont was trying to welsh on a deal?

Newmont made the sale at $1.56, begrudgingly after careful consideration. You could hear the teeth grinding from its office in Perth. So Eric Sprott did the second-best deal he had ever done on Novo shares. Like he needed the money. But on the other hand, principles are principles, especially when money is involved.

The really big news from Novo, and the first real indication of the potential size of the gold deposit, came on July 12, 2017. Quinton wrote the press release [54] while he was on a trip to Japan for Irving Resources. The first paragraph included the following. The emphasis is mine.

> Novo [. . .] is pleased to announce that it has found *in situ* gold nuggets up to 4cm long in primary conglomerates from its first trench at its Purdy's Reward

prospect and has collected a bulk sample of these gold-bearing conglomerates for analytical test work. The sample originates **from a one meter thick reef near the top of an 11 meter thick stacked sequence of mineralized conglomerate horizons**.

The whole exploration team was present when taking the sample, and there is a video[55] of Barry Rattingan using a portable jackhammer to pry out a great big nugget. Brad Smith had been using the metal detector. He sees the nugget when they turn over the rock and says, "Oh, shit." The jackhammer belonged to YT. He and Brad Smith provided the expertise with the metal detector. Luke Meter and Allan Ronk provided the geological thinking.

They found good gold. The sample was of a reasonable size, at seven hundred kilograms: two meters by two meters, and 5.2 cm thick. The metal detector was used first, to find an area giving a strong signal.

The sample was split into two and sent off to the lab in Perth. But first, Novo ran the material through a Steinert mineral sorting machine. Even at the start Quinton was thinking about how to reduce costs. The sorting machine sorted the first sample to a total mass of only 2.15 percent of the original material. The second sample came in at a tiny 1.82 percent of the original mass weight. After sorting, the material was sent for assay.

Costs in mining are always based on how much material you are working with. If you could get rid of 98 percent of the weight with an inexpensive sorting machine, your total processing cost would really decline.

Quinton's press release included a link to that short video. To the best of my knowledge, it was the first time a video had been included in a press release about finding gold. Since it would be easy to fake results in such circumstances, it surprised me that the exchange permitted it. But investors evidently loved the video as the share price more than doubled, going from $1.49 when it was released to $3.15 on the day when the assay results were published.

Those results arrived on August 8, 2017. One sample graded 87.76 grams of gold per ton (g/t), and the other 46.14 g/t. The average grade of the two samples was 67.08 g/t, and this was the most meaningful figure, as the two assays were from the same sample, just split.

When the results were released, the stock continued to rocket higher. In the first week of October, Novo shares peaked at $8.83, up an incredible 1,700 percent since starting the year at forty-eight cents.

There was one slight technical problem, actually two problems, that could and should have been avoided.

Many people in the peanut gallery [56] have been tossing rocks at Quinton for years, mostly through jealousy. The real issue would cause confusion and anger among shareholders who ended up buying at the top and watching the share price tumble for years.

Above, I highlighted an important sentence in the press release of July 12: "*The sample originates from a one meter thick reef near the top of an 11 meter thick stacked sequence of mineralized conglomerate horizons.*"

But it didn't. It came from the bottom.

This wasn't Quinton's mistake. It was a mistake by the geologists on the ground. Quinton was in Japan at the time and didn't catch it; really, he couldn't have at the time. But when the assays came out a month later he should have caught it. He didn't. It remains the only giant mistake I have ever seen Quinton make in all the time I have known him.

A second press release,[57] on August 8, 2017, repeats the error: "*As discussed in Novo's news release dated July 12, 2017, this sample originates from the uppermost horizon of an 11-meter thick sequence of mineralized conglomerate beds.*"

It would take months of ground work and drilling to determine the actual structure of the conglomerate. Eventually Novo realized that the sample had not been taken from the top of an eleven-meter sequence. It was the difference between day and night. The sample came from the bottom of the reef, not the top.

I am the only person so far, writing about Novo, to point out that in any conglomerate sequence, since gold has a specific gravity of about 19 (not being pure), as water flows over the gravel the densest material will sink to the bottom. The richest quantity of gold is always at the bottom of any conglomerate sequence. The error in the press releases made it look as if this was the richest reef ever found. It wasn't. Everyone who believed it was alluvial gold in a reef supposed it would get richer and richer as they went lower in the sequence.

Quinton wasn't on site; he was in Japan when both press releases were issued. But in the Navy, when a ship runs aground, the captain is always court martialed.

This brings me to the second howler.

I return to the Bre-X scandal of 1997. It was a gold project in Indonesia. The geologist salted the samples with shavings of gold from his wedding band at first, and later by buying from the locals some sixty-one kilograms of alluvial gold. He fed small bits of gold into the drill samples being sent to the lab for assay. The project grew from a two million ounce resource to an incredible seventy-one million ounces in eighteen months.

No other gold deposit in history had grown so fast. And neither had the Bre-X project, because it was all fraud, with gold in every drillhole. Literally, it was too good to be true.

When Novo's assay results came out on August 8, 2017 someone should have noticed that there could not possibly be 67 g/t gold in nuggets *at the top* of a conglomerate sequence.

There was one more matter. No one else has ever mentioned it, but it too was and is significant.

They used a metal detector to determine where to take the sample. Of course it was going to be high-grade; it had nuggets in it. That's how metal detectors work.

But miners don't give a shit about cherry-picked samples. They have to have representative samples.

All the work done since 2017 has shown that the grade at Purdy's Reward and Comet Well was 2+ grams of gold per ton over perhaps two meters. The gold is found in the bottom of the sequence, not the top. The sample was not taken from the top of a reef but from the bottom, and it did not represent the average grade of the reef. Much of the reef was correctly barren.

So it wasn't eleven meters of high-grade, with a likelihood of even richer grades deeper in the conglomerate. It was a couple of meters of lower grade.

The mistake was in causing investors to believe the deposit was far richer than it was. You may safely assume Kirkland Lake were pissed after figuring out it wasn't as rich as they thought. They made the same assumption Quinton did, and frankly, it was wrong.

The saying "Only mad dogs and Englishmen go out in the midday sun" originated in the Pilbara region, not in India, as has popularly been supposed. There is crazy, and there is Pilbara crazy. Novo was entering a three-year period in which the nutcases flourished. Pilbara crazy became the rule, not the exception.

Until Brent Cook wrote his first piece on the project in mid-

2017, just as things got interesting, the only newsletter writers who covered Novo were Jay Taylor and me. All of a sudden, the rush was on to claim the discovery of the Pilbara.

John Kaiser swept into the space as if the industry was sitting waiting for his wit and wisdom. According to him, the Novo holdings might contain as much as twenty billion ounces of gold. He did sort of forget to mention that most of his information came from Jay Taylor and me.

I found that to be an interesting number, firstly since it was pretty much based on just two assays, and secondly because only about seven billion ounces has been extracted from the Earth's crust since Abel took his gold pan down to a stream in the Garden of Eden. When, months later, it became obvious that while there was a lot of ground with conglomerate containing gold nuggets, Kaiser said he didn't know if two grams of gold per ton at the surface was economic. That was just as dumb as the projection of twenty billion ounces.

There was an even bigger fruitcake in the pantry named Allan Barry, or Allan Barry Laboucan, depending on the time of day. I had a run-in with him a couple of years earlier. I posted a piece saying I believed we were about to go into a market correction. He came out and attacked me on chat boards, saying I had no idea of what I was talking about, and was wrong.

I wrote him back and suggested that he give it a month to see what transpired. Logically, he couldn't say I was wrong until we waited to see if I was in fact wrong or not. I asked him if he would be similarly quick to apologize if I turned out to be right, and there was a correction.

There was a correction; I was right. He never bothered to admit being wrong in what he said. He was one of those guys who has an opinion about everything, runs his mouth, but can't admit it when he is dead wrong.

He liked David Lenigas a lot, and rather than focus on Novo, Barry sucked up to Lenigas in the hopes of getting him to subscribe to his minor website with its few followers. He also runs a sub-tiny junior, and once issued a press release saying it had drilled into a vein of quartz, and that in that area, gold was sometimes found in quartz. That's perfectly true, but his quartz vein did not contain gold, just as most quartz veins do not contain gold.

Neither Barry nor Kaiser had any real insight into the Pilbara or Novo Resources. They just wanted to pretend they had discovered the company and the project. They didn't do either, and soon faded away into well deserved obscurity.

CHAPTER 17
THERE IS CRAZY AND THERE IS PILBARA CRAZY

ON MAY 26, 2017 NOVO ANNOUNCED a binding letter of agreement with Artemis, adding another 1,536 square kilometers to its already giant land position, now amounting to 6,021 square kilometers in addition to Comet Well. Novo took the first bulk sample from Purdy's Reward on July 12, 2017. Since a metal detector was used they knew the results would be good, perhaps even excellent.

I was married to a Brit for thirty years. She wanted to make it to her seventy-fifth birthday. She came within a week of doing so, but close only counts in horseshoes and hand grenades. I miss Barbara greatly; she was a wonderful wife, mother, and grandmother.

We would argue regularly about the meaning of words, as British and American versions of English often diverge. I think savory means tasty; she thought it meant the opposite of sweet. She spelled it differently, too. Any time you have people from different countries using the same language, there may be fundamental disagreements as to what certain words mean.

David Lenigas signed the binding letter of agreement with Novo in May of 2017. Once the rich sample loaded with gold nuggets was taken from Purdy's Reward, Quinton knew he had to drag Lenigas to the table to sign the definitive final agreement. He chased him by phone and scheduled meetings and tried every way he could to get Lenigas nailed to the floor. Lenigas came up with one excuse after another and canceled one meeting after another. For Quinton it was like trying to pick up a drop of mercury with a pair of tweezers.

The problem was that Lenigas thought of himself as a pirate, literally. He believed that being a robber baron was a great job, for which he was highly experienced. His definition of the word "binding" wasn't in any dictionary. As far as he was concerned, once you make an agreement and sign on the dotted line, that's when you start to negotiate.

Lenigas pulled the slickest piece of legerdemain I have ever seen any conman pull. Quinton was in a bind of his own making. He had an agreement with a pirate who wanted to pump up the value of the twenty-five million shares of Artemis he had been handed for being such a nice guy. Yo-ho-ho, and a bottle of rum.

Quinton had the assays from the sample taken in July. He couldn't have them in his hands and not release them. What he did not have was a definitive and absolutely binding agreement with Artemis.

So Lenigas paid himself, paid his backers, and screwed his own shareholders. On August 4, shortly before Novo would release the bulk sample assays, he announced a private placement for 23.7 million shares at 12.66 cents apiece. That was about $3 million. It all went to Lenigas and his buddies.

Once the first bulk sample results from Karratha were released, on August 8, Lenigas had Novo by the short and curlies. He started going on the HotCopper chat board in Australia to let Quinton know who had the power. Here are five posts, dated August 10 and 15.

> werdna007 [58]
> 10/08/17; 8:25; Post # 26405439 [59]
> Hi Doodledog,
> I agree with your explanation, well done.
> There is no relationship between ARV and NOVO, at present, but subject to certain conditions being satisfied,

there may be in the future. Ed Mead, as a senior geologist and competent person of ARV, would have to be satisfied with the validity of all the statements made by NOVO in that announcement. ASX listing rules must be observed for ARV to release an announcement.
Cheers

werdna007
15/08/17; 17:19; Post #: 26512020 [60]
ARV own 100% of Purdy's. NOVO own 0% at the moment. NOVO's MC has risen more than $400m on ARV's tenement.
Stay tuned.

https://hotcopper.com.au/search/search?type=post&users=werdna007
My estimate of NOVO SP value
werdna007
15/08/17; 18:24; Post #: 26513184 [61]
20–40c, maybe 50c with overvalued Beaton's Creek.
Very little workable real estate for quite some time and little cash to do anything more than that.
No mill to process gold...no native title...no exploration licences....no mining licences.....no money...... no FIRB approval...no TSX approval....no ASX approval...no environmental approvals.....no understanding of the MOA they entered into....almost says it all.

Cheers

werdna007
15/08/17; 19:28; Post #: 26514402 [62]
ARV has NOVO by the short and curlies!

My estimate of NOVO SP value, page-8
werdna007
15/08/17; 22:05; Post #: 26516802 [63]

ZEN, have reconsidered your request and have placed absolute minimum MC's on a few of ARV's tenements over the course of time – minimum values only on (1), (2) and (3)

(1) Purdy's - $1 b mc - wits gold

(2) Oscar - $1b mc - wits gold

(3) Carlow Castle - $1b mc - Cobalt-gold copper

(4) Mt Clement - JV with Black rock - gold mining - $250m mc

(5) other - diamonds - platinum group metals- antimony - whundo copper- silica hills - $100m

(6) Radio Hill Plant refurbished and other fox jorced tenements - $150m

Total $3.5B with 440m shares on expanded capital as at 30/9/17 equals approx $8.00 per share.

With confirmation of wits gold at purdy's and oscar (over the next 6 mths to 3 years) an increase of mc of $108.5B + Carlow Castle (with high grade cobalt/gold/copper) an increase in mc of $20 B to a combined share price of $300.00 per share minimum within 3–5 years (on capital of 440m shares) - profit taking opportunities at $50/$100/$150/$200/250 and $300 a share over that period for around 10–16% of shares per segment.

Please note that all these calculations are based on ideal outcomes and the actual outcomes may vary greatly from these outcomes.

Please also note that these possible outcomes are all based on my own calculations and are not in any way indicative of actual outcomes.

Please also note that these outcomes are based on my own private calculations into the future and the actual results may be wildly different from these.
Please also note that I am just a private investor and none of my forecasts may actually take place.
I am not a paid analyst, nor associated with any Company or Stockbroking firm or any party who has inside information or advance notification of price movements or corporate activity in the field.
I am just like you, who spends 50 hours or more of research on stocks I would like to keep or trade, and will ,in the future, keep or trade , ARV stock at any time I see fit, but not at any time below $10 per share and mostly not below $50 per share, if that ever happens.

Cheers to all prospective SH and traders.
May we all achieve our objectives for massive profit or comfortable living.

For the chairman of a publicly traded company to post something like Lenigas was posting on Hot Copper, under an alias, with a private placement pending and while working with Novo on a definitive agreement on Purdy's Reward, is way beyond shithouse rat crazy, and beyond 10 on the Pilbara Crazy scale. It's totally bonkers and symptomatic of an utter sociopath.

Novo was selling for $4.85 a share on August 15, 2017 when Lenigas posted his estimate that it was worth only 20–40 cents a share, or maybe fifty cents. With all the ups and downs since then, Novo shares are selling for $2.20 as I write this.

But what of Artemis? Artemis was selling for $.215 a share when Lenigas was posting his prediction of $300 minimum within 3–5 years. Artemis is now $.135 a share, and three and a half years have passed. Artemis will have to get cranking soon, if it is to reach even that first profit-taking opportunity, of $50. The share price would have to increase by a percentage in the tens of thousands in the next eighteen months. The $300 fantasy share price is far less likely still.

Looking back, the best part of the prediction from Lenigas was his statement that the Radio Hill rust plant was worth $150 million. How about $1.50 total for the plant? Artemis has been promising for years that it is on the verge of production.

It goes without saying that Lenigas was breaking more rules on full disclosure than I can count. The Australian Securities Exchange knew something funny was going on. They sent him a letter, demanding to know why Artemis did a private placement for insiders when it could reasonably expect the amended earn-in agreement to have a material effect on the price of the stock.

Novo did get Artemis to the table to sign an amended earn-in agreement on August 15, 2017. Lenigas demanded and got an additional four million shares of Novo worth right at $20 million because, as he had posted earlier that day, *"ARV has NOVO by the short and curlies!"*

All the Australian posters on Hot Copper thought Lenigas had pulled the coup of the century. After all, who wouldn't want to screw their partner at the first opportunity? I don't remember anyone suggesting that it might have been a nice chunk of change short-term, but in the long-term might not be such a great deal. I looked at it as taking a hooker with you on your honeymoon. While interesting and maybe even fun, it would tend to set the tone for the marriage.

The ASX had a fit, and on August 16, 2017 made its demand for an explanation about the timing of the private placement on August 4, the assay results on August 8, and the corn holing of Novo on August 15 under the entirely different terms for the "binding letter of agreement."

Lenigas was as cute as a cartload of monkeys when he responded to the ASX on August 21. "Artemis was not aware of the Amended Earn-In Agreement prior to conducting the Placement. . . " And, of course, that was perfectly true. Lenigas may well have known he intended to screw Novo, but he hadn't told them yet. He would do the placement knowing full well that the assays were about to become known, and as he said on Hot Copper, "*ARV has NOVO by the short and curlies!*"

But he didn't know on August 4 what the terms would be, so he didn't really lie through his teeth to the ASX. Not really.

Earlier in this book I talked about the games people play. In this particular game Lenigas and his shareholders thought he was playing a zero sum game, where Artemis extracted $20 million worth of shares from Novo, and Novo lost $20 million in value in exchange for Lenigas redefining the term "binding". So Artemis seemed to have won and Novo seemed to have lost.

No one realized it at the time, certainly not the Artemis shareholders. While it was a great deal for Lenigas because all he wanted was to pump up the value of his twenty-five million shares so he could dump them, he screwed his own shareholders and Novo's. The reason is easy to understand.

Novo had 50 percent of Purdy's Reward, after handing over that $20 million in shares. It had 80 percent of Comet Well, that it had already paid for. Novo also had hundreds of square miles of ground surrounding Purdy's Reward and Comet Well, and owned that 100 percent.

The only reason Novo wanted Purdy's Reward was because

Purdy's had permission to drill. In time, permissions would follow for Comet Well and the rest of Novo's ground, but Novo could commence work immediately at Purdy's Reward. Novo didn't really need Purdy's. Quinton needed the right to drill and carry out exploration. He could start at once at Purdy's.

So, as time passed and Novo worked at Purdy's Reward, the question would naturally arise: how did Novo intend to spend its treasury once it had a drill permit for Comet Well or anywhere else on its land around the Karratha area? Would it be spent on the ground it owned 100 percent, or on Comet Well (80 percent owned), or should it continue to pour money into Purdy's Reward (50 percent owned), where Lenigas had just shown them that he would screw them at every opportunity?

At Purdy's Reward, there was even permission to take a bulk sample. That was highly valuable from a technical and geological point of view. It would provide a lot of valuable information. I think it was for twenty thousand tons. For some strange reason Novo didn't take advantage of it. When the drill permit arrived for Comet Well, Novo moved all its equipment there, and left Purdy's Reward to produce dust that could blow away when the winds came up.

It would take a couple of years for a few brave souls on the Hot Copper board to start to ask why Novo wasn't spending its money on Purdy's Reward when it could have. They should have checked the thermometer, because it would be a cold day in Hell before Novo would spend another dime there as long as Artemis owned half of it. Lenigas not only stiffed Novo, he stiffed his own shareholders.

By why should he care? He had all the money from the free shares he had been given.

On August 23, 2017 I wrote a piece blasting Lenigas. While everyone realized he had screwed Novo, few understood that he had stiffed his own shareholders as well. That realization would come later.

In that piece I tried to make the point that Lenigas was shitting in his own lunch bucket. Artemis was partnered with a fully cashed-up Novo and was getting the services of the best geologist in the world, in my opinion. The first thing Lenigas does is to steal from him, for his own benefit.

But at that time I didn't understand his motivation. I didn't know that he was sitting on millions of shares he wanted to blow out as quick as he could, so he could mosey on down the trail to some other scam.

Here is my conclusion to my August 23 piece: [64]

Let's pretend Novo's shares shoot up when the original video came out on YouTube on July 11th and Lenigas puts out a press release from Artemis. "Congratulations to our JV partner with their success at our 50% owned Purdy's Reward. Artemis is thrilled to have gained Novo Resources as a valued associate in this venture. Artemis is overwhelmed to have joined the team of Newmont Mining 3rd largest gold mining company in the world, Sumitomo Corporation, Mark Creasy, the most successful prospector in Australian history and Eric Sprott of Canada, billionaire investor and we get the services of one of the top geologists in the world for free and two million dollars spent advancing our projects. Here's my pen, where do I sign the definitive agreement?"

What do you think the shares would have done that day? Well, they wouldn't have gone to the mythical $8 a share but they would have easily spiked to $1 a share.

Instead, Lenigas gets cute, cuts a deal good only for him and his eastern European backers and stiffs both Novo and his other shareholders.

While Lenigas fully understood what he was saying when he kept issuing the veiled threats to Novo in a number of press releases, he needs to pay a lot of attention to the shot the Australian exchange just fired across his bow. He's playing with the big boys now and the chicken shit little games need to stop at once.

This was not the first time I had embarrassed Quinton; nor would it be the last. Naturally Novo had to issue a press release suggesting I was totally off base. It began as follows:

Novo Resources Corp. [. . .] announces that it has learned of certain recent public media disparaging one of the Company's joint venture partners, Artemis Resources Limited ("Artemis"), and its Chairman, Mr. David Lenigas. Novo wishes to make clear that it does not condone such views.

David Lenigas had just robbed Novo shareholders of $20 million Canadian in shares. He had ensured that Novo would never trust Artemis any further than they could heave him, but you aren't allowed to say such things in a press release. Actually, I think that's exactly what press releases should say, if it's true.

Shortly thereafter Novo announced an investment in Novo by Kirkland Lake at $4 a share, to the tune of fourteen million shares for a total of $56 million, all based on the extraordinary assay results released in early July. It would be the start of an uneasy relationship between the two companies. Novo was looking for a fat treasury. Kirkland Lake wanted a cheap entry into a potential takeover target. As placements go it was rich, at $56 million, and generous with the shares, plus another fourteen million warrants at $6 apiece to Kirkland Lake.

In September of 2017 Novo began serious exploration at Purdy's Reward. The biggest gold show of the year, the invitation-only Denver Gold Forum, was scheduled for later that month. Quinton and Novo were determined to make a splash. They did, but in time it came to look more like a belly flop. Years later, Quinton would suggest that it was probably a mistake to have done it. But at the time it was really cool, and unlike any presentation any of the viewers had ever watched before.

Novo was given a fifteen-minute slot at the Forum, at a time that would be in the middle of the night in Karratha in Western Australia. Novo rigged spotlights and used heavy equipment to clear off a good spot to take a sample. Again, in hindsight, using a metal detector to find gold will only finds gold where it beeps; it doesn't tell you how representative the sample is.

The July assays had created unrealistic expectations. The Denver Gold Forum presentation reinforced those unrealistic expectations. It was a mistake that would take Novo years to get over.

Quinton introduced the company and the project to the watchers at the gold show before handing over to Luke Meter in Karratha; he was Novo's exploration manager at the time. Luke spent a couple of minutes talking about their first trench, where they had marked out gold nuggets found with a metal detector with pink spray paint. There were a lot of nuggets and it had to be impressive. He handed over to Brad Smith, who I mentioned earlier as one of the three owners of Comet Well, and whom Johnathon Campbell had brought in to prospect his SPL at Comet Well.

Brad Smith fires up his metal detector on the video and begins to move around the 30 × 20-meter trench, showing one nugget after another. Brad tended to be a Chatty Cathy and could have shown the same thing in a quarter of the time. I'm not sure what he thought he was supposed to be doing, but in essence he was lecturing on the use of a metal detector.

Finally he hands back to Luke, who had two of his guys prise out some marked nuggets with a jackhammer. John hammers out a tiny melon seed nugget and it comes free from the rock, with Brad working the metal detector.

The Denver Gold Forum is unique in that it is run extremely professionally. Your slot runs for exactly fifteen minutes and not a second more. They will cut your mike off and go on to the next presenter. Quinton was standing at the podium, watching the action in WA and sneaking regular glances at his watch. It looked like Novo's presentation was going to run over time. Finally, Pete and Brad pull out a good size nugget and show it to a suitably impressed crowd in Denver.

What no one understood at the time, in either Colorado or at Purdy's Reward, was that if you clean the conglomerate level with an excavator down to where the nuggets are, you will find a lot of nuggets. But unlike the advertised eleven meters of conglomerate, it was a lot more like two meters containing all the nuggets. Instead of the sixty-seven grams of gold per tonne that was hoped for, it was a lot more like two grams.

It was a mistake, and one that would cost Quinton Hennigh and Novo a lot of credibility. The mistake was in initially assuming they were testing the top of the conglomerate section, when in fact they were testing the bottom, where the richest gold had to be.

The way in which the gold industry tends to work is that everyone makes mistakes, but also everyone wants to cover their mistakes like a cat in a litter box. The difference between two grams and sixty-seven grams of gold per tonne was so great that it demanded an explanation, but the market didn't get it. So the Novo naysayers beat up on the company for years.

In my view, someone should have come out and said, "We screwed up. We don't have sixty-seven gram rock but we do have hundreds of square miles of two-gram material at or near surface."

CHAPTER 18
THE PEAK, AND THEN THE LONG DECLINE

ON THE STRENGTH OF THE INITIAL ASSAYS from Purdy's Reward, the injection of $56 million by Kirkland Lake Gold, and the video show at the Denver Gold Forum, Novo's share price peaked in the fall of 2017. It was a heady time for the company, rising from twenty cents a share in 2010 to $8.83 on October 1, 2017.

A return of 4,300 percent is what every junior resource investor hopes to find once or twice in a lifetime. But holding onto that kind of return can be difficult. For the next three years Novo would struggle to regain credibility, with a cast of thousands of keyboard commandos in the peanut gallery throwing rocks with every new press release.

Shares of Artemis would peak one month later, in November, at just over forty Aussie cents, before starting a long and painful decline to a low of $.025 by December of 2019. When Lenigas signed the original "binding" letter of agreement with Novo in March of 2017, Artemis shares were about eleven cents. When he got his twenty-five million shares in late 2016, they were worth two cents apiece. They went up twenty-fold and came right back down to where they started.

I can't see exactly where Lenigas dumped his shares. The Artemis annual report in June of 2017 shows him owning all twenty-five million. A year later there is no record of him owning any reportable shares. The shareholders didn't realize it but he had pocketed something up to $10 million for a year's work. Once he unloaded his shares, he just coasted. He had accomplished what he wanted to accomplish.

One important issue that I don't recall anyone mentioning

about the conglomerate find at Karratha was just how unusual it was. The deposit was virtually identical to that of a high-grade alluvial gold project, but since it had been covered with basalt for 2.7 billion years it was now hard rock — hard rock with the characteristics of an alluvial deposit. Beatons Creek was similar in nature but it contained loosely consolidated material that was perfectly suitable for gravity recovery.

For the time being, Beatons Creek and Nullagine were on the back burner. Millennium Minerals had never made a profit and clearly never would, but kept limping along. The price of gold was tending higher and the Aussie dollar was declining, and that was enough to keep the corpse warm.

Novo had the easy-to-mill gold. Millennium had a mill and a lot of uneconomic rock. Millennium wouldn't consider doing any sort of deal with Novo, even if it might keep Millennium running for a bit longer. That was dumb but you can't fix stupid.

Quinton was just about the only person in mining who understood that much of the gold in the Pilbara came from the conglomerate reefs. Beatons Creek was pretty close to being conventional in nature, with smaller gold, which was at least somewhat measurable in its quantity and grade in the reef. Purdy's Reward and Comet Well were another kettle of fish.

In Quinton's first attempt to measure the gold, he used a large-diameter drill rig with a large bore for digging a big hole. It was designed to drill water wells. His theory was that if he picked up enough material he could get a reasonable idea of grade.

At the same time, Novo's geological team used a diamond core rig to determine the structure of the reef. It worked great, and provided a clear three-dimensional picture of the reef. By November of 2017 Novo had completed over sixty short core holes indicating the structure of the conglomerate reef.

The large-bore water drilling rig, on the other hand, was an abject failure. It would grind up the material, but since gold is about nineteen times heavier than water, any nuggets in the hole just dropped to the bottom. You could pull the drill out and vacuum out the cuttings and nuggets, but you had no idea where in the reef they came from. The reefs ranged from a couple of meters thick up to twenty meters, so knowing where the gold was would be key to mining at a profit.

For the first time, Novo acknowledged that most of the nuggets found by the metal detectors were located near the bottom of the conglomerate sequence. If the geologists had ignored the fact that they were dealing with hard rock and pretended it was nothing more than a gravel bed, similar to any alluvial project, it would have been obvious to them from the start that the lower part of the sequence would be the rich area.

That's a failing throughout the gold mining industry; geologists trained for hard rock exploration look down on placer miners and don't want to learn what makes those deposits, or their nature. Placer miners know placer deposits, and hard rock miners know hard rock deposits. Rarely do they mix. That's just dumb.

To obtain a mining permit in Australia it is necessary to prove to the mining department the grade and quantity of the gold. But in a large nugget system, such as Novo faced, it was almost impossible to measure to government standards. Novo couldn't use a conventional core or RC rig, as the nuggets were too big and too widely spread out. The water well drill rig left all the gold at the bottom of the hole.

Figuratively speaking, since not a great deal of it remained, Quinton was pulling his hair out. Without measuring the gold to Australian standards he would never get a mining permit.

My attitude, as I said a number of times when I wrote about

Novo, was that you can't measure nugget gold; you can only mine it. The Romans wouldn't have had any problem mining the gold-bearing reefs. The Spanish would have grabbed their picks and started digging. Hell, the *garimpeiro* miners in Brazil today would have mixed some diesel fuel and ammonium nitrate and started blasting holes in the reef to pluck out gold nuggets until they made enough money to feed their family. None of them would ever get permitted in Australia.

Quinton and I talked dozens of times. I kept repeating that you can't measure nuggety gold, you have to mine it. He insisted the Department of Mines would never permit it until it had been measured. In the end he began to take bulk samples to determine grade. As far as I can tell, taking a bulk sample is the exact same thing as mining.

I cannot over-emphasize how difficult a project the Karratha conglomerate gold story was. There was one simple reason why over a hundred and thirty years had gone by and tens of thousands of miners had walked right past the gold in the conglomerates. They didn't understand it and couldn't figure out how to mine it. Quinton understood it but hadn't quite gotten around to figuring out how to mine it.

One Novo decision from early in the program was to make use of sorting machines. Quinton believed he would find small, finely disseminated gold that could easily be drilled to determine grade and quantity. It would never happen.

He chanced on the idea of the sorting machines. They can sort just about anything, based on color or size or density or material. They are commonly used to sort trash, especially trash containing metal. Their use is skyrocketing in the mining industry. The technology is advancing rapidly. Right from the gitgo, Quinton included representatives from two companies making the sorting machines. They will revolutionize the industry in time.

Australia depends on the mining of coal and iron ore for a major portion of the taxes paid to the state and federal governments. With billions pouring in from China each year, Aussie miners are the highest-paid miners in the world. A truck driver can make $200,000 a year. The salaries are amazing. If you visit Perth airport, all you will see is thousands of miners waiting to fly off to a mine site or just returning. There seems always to be a labor shortage in the country.

As a result, the crews can be picky. Australia is a socialistic country. The workers believe everything should revolve around what they want. As such, starting up a major mining project is similar to herding cats. The workers go where they want, when they want. Management is fairly low on the power pole.

Beatons Creek had been held in abeyance since it made no sense to expand the resource further. Quinton had outlined a resource. He had done test mining and processing and was now stuck until Millennium could be budged from TDC. Beatons Creek was important to Novo but was clearly not a company-making property unless and until it could be put into production. But the Karratha collection of projects, taken together, clearly were a company-making proposition.

Quinton had brought on Leo Karabelas in 2011 to handle corporate communication for Novo. Leo would do the same thing for a whole slew of companies with a Hennigh connection. Bringing on Rob Humphryson as CEO in June of 2017 was a major achievement. As in every other country, Australians tend to prefer working with their own kind. No Canadian company was going to march into Australia and start telling them how the mining game works. Rob took a lot of the burden off Quinton, and ever since the COVID-19 nonsense began he has had to shoulder most of the work, keeping exploration moving to get into production.

The exploration staff at Purdy's Reward and the Karratha area was more problematical. Everyone had their own idea of how the exploration should be conducted. Getting everyone to march in the same direction was difficult, to put it diplomatically. People came and went. One senior fellow simply didn't show up one day, or ever again; he waltzed off and no one had any idea of where he was or what his problem was.

The management team was excellent, and worked smoothly together. Ronan Sabo-Walsh kept track of the money and did an excellent job of it.

Since the hard rock conglomerate deposit was close to being unique, no one had any experience of dealing with it. Over time, this showed. There were a lot of mistakes made and then instantly pointed out on various chat boards.

People afraid of making mistakes will never succeed at anything. Everything is a process of trial and error. Even flight manuals approved by the FAA will talk about straight and level flight as a series of turns and banks, with both climbs and glides around a central direction. If you aren't making mistakes, you aren't making enough decisions. That said, while it's only human to err, if the erasers on your pencils always wear out before the lead, God is trying to tell you something.

As the chat room warriors noticed immediately, Quinton made a number of mistakes in 2017 and 2018. But it's OK to make mistakes when you are doing something for the first time; you are bound to do so. It's OK as long as you don't make the same mistake again and again. Quinton made mistakes but rarely made the same mistake twice.

Late 2017 was an exciting time, as major progress looked to be under way. As a major shareholder, Eric Sprott joined the board in November. Greg Gibson was already a board member.

A heritage agreement was penned with the local Ngarluma Aboriginal Corporation. That was necessary, to get permission to begin exploration at Comet Well and the other surrounding tenements. Until Novo could explore and drill Comet Well, Artemis had the company by the short and curlies, as David Lenigas so kindly put it, as Novo was obliged to work at Purdy's Reward.

Artemis did announce that the Department of Mines had granted a permit to take a 20,000-ton bulk sample at Purdy's Reward. If anyone other than David Lenigas had been running Artemis, that permit would have been valuable to both companies, because for the first time a real mining sample could have been taken. That would have saved a year of fits and starts and disappointed shareholders demanding instant gratification. But anyone who has ever been bitten by a rattlesnake knows you don't put your hand down to pet the next one to wander by.

By mid-December 2017 the Department of Mines had granted the tenement for Comet Well to Novo, which meant exploration could begin. In late December Novo issued a press release talking about results at Purdy's Reward, where some sixty-nine diamond core holes had outlined the conglomerate reef across a width of five hundred meters and a length of over fifteen hundred meters. By now Novo had picked up over 12,000 square kilometers or over 4,600 square miles of property. Novo controlled one of the largest land positions in Western Australia.

Trench samples from Purdy's came in at 15.7 and 17.7 grams of gold per ton, from bulk samples of over three hundred kilograms. Novo was realizing that to obtain an accurate estimate of the gold grade, the sample size had to be larger, perhaps much larger. There remained the question of where in the reef the gold came from. If you were to test only the lowest part of the reef, you would get exceptional grade. If testing the entire conglomerate sequence, it would be much lower.

Kas De Luca joined Novo in early 2018 as exploration manager, from Newcrest. Getting everyone on the exploration team to work in the same direction was still difficult. She was an important addition to the Novo team.

CHAPTER 19
NOVO SET THE BAR TOO HIGH AT KARRATHA

THE DISCOVERY OF GOLD in the conglomerates in the Karratha area by Quinton and his team was a sort of good news–bad news story. Thousands of prospectors and miners had worked the region for over a century. Weekend warriors with metal detectors took millions of dollars' worth of gold nuggets out of the area for several years. Until Quinton came along, no one put it all together. Dozens of tiny Australian junior mining companies were all searching for the mother lode; that is to say, lode gold. Quinton didn't care, thinking that after all, gold is gold. Who cares whether it is in hard rock vein systems or in nuggets in the conglomerates?

If you want to make a monster discovery — and that was the first item on Quinton's to-do list — you must start with a good theory. His theory of gold precipitation out of brine and salt water 2.7 or 2.8 billion years ago inferred a lot of gold. It just didn't suggest where exactly it was.

Beatons Creek was actually a sort of slam-dunk. The project was close to being conventional. While the nugget effect made the gold hard to measure with accuracy, that same feature made it easy to process. Although the gold in the conglomerates had long been known about and mined, at least on a small scale, without the theory in hand, miners didn't realize that where there were conglomerates, there was gold. Since this sort of deposit is almost unique in the world, it was overlooked by prospectors and the management of junior gold lottery tickets in their search for lode gold.

Beatons Creek was so close to conventional that it was no stretch for Quinton to put together a feasible project. There was the one small issue of having no suitable mill, but Millennium was trying hard to go out of business, after which Novo could use their mill.

The grade of the first two samples from Purdy's Reward set shareholders' expectations so high that Quinton was under the gun for years, both from several respected geologists writing about the company and from a bunch of clowns writing about a project they had never visited and didn't understand. It can't be fun standing there dodging cobbles all day.

Nobody in history ever discovered gold by finding an exceptionally high-grade sample at the first attempt. Whatever you come up with, some samples will be higher, most lower. But by announcing two samples averaging sixty-seven grams per ton, Novo elevated the bar right to the moon. Stating incorrectly that the sample was taken from the top of an eleven-meter segment of reef just made it worse.

Sampling and small bulk sampling continued into 2018, with work being done at Purdy's Reward and, for the first time, at Comet Well. Due to the nugget effect, no two samples ever showed similar grades. At Purdy's, Novo mapped the conglomerate structure very well with a large number of short core holes. Bulk samples from the conglomerate showed gold nuggets over a two-meter interval, in the range of two to three grams per ton, all located near the bottom of the gravel sequence. That would be economic anywhere in the world. With Novo controlling around 12,000 square kilometers of land in the Karratha area, that is potentially a lot of gold. But at Comet Well there were actually several reefs, all near each other, all carrying various grades of gold.

You can't mine it until you have measured it, and you can't measure it.

I thought the whole issue was stupid. The gold was

economic and near surface. This discovery would have been put right into production at any point in history prior to the Bre-X scam. So in the name of protecting investors, more regulations and rules would somehow make investing safer. But this was Australia. Welcome to modern professional mining.

Throughout 2018, more and more samples were sent to the lab for assay. Some personnel issues popped up, with blame due on both sides, and with the net effect that results took far longer to be published than impatient investors would tolerate. In my mind the issue went all the way back to those first two assays. Investors couldn't come to grips with the fact that those results simply were not reflective of the nature of the deposit.

Using a metal detector is a great way to find nuggets but a terrible way to sample areas. You can find gold but you could never mine it that way. Of course the samples will grade high. But those samples do not reflect anything other than the tiny area they came from. The initial samples went only centimeters below surface. They didn't mean anything in terms of the overall project.

Communication would have helped, but many of the problems became obvious only well after the fact. There is a lot of trial and error in any mining project, and this was the most unusual gold deposit in the world. Karratha provided a sufficient supply of trials and a selection of errors.

Quinton was working on a couple of problems at the same time. It was obvious to the market that the use of mechanical sorting machines was a high priority for him, since the first discovery at Purdy's Reward. Novo would spend the next three years working with the technical teams of two companies to find a way to process crushed rock to concentrate the gold in the smallest amount possible. I've been to Perth and seen such a machine in action. When a machine can pick up pinhead-size gold and separate it from the gunge, that's quite remarkable. Novo has been able to reduce the total mass down to less than one percent of the volume it starts with.

At Artemis, David Lenigas had managed to dump all his shares and was busy blowing the remainder of the company's cash. Artemis still had the four million Novo shares he had extracted in mid-2017. In May of 2018 Kirkland Lake agreed to buy them, to add to the twenty-one million it already owned. When Novo issued the shares to Artemis in August of 2017, one condition of the deal was a twelve-month hold. Novo agreed to release Artemis and Kirkland Lake from that condition. Kirkland Lake paid $5 a share and sent the money to Artemis in May.

The transaction made Quinton slightly nervous. Whenever any party becomes a large shareholder in another company, they tend to want to start making decisions for the company. They feel they should have authority but don't want the responsibility. It's always handy to have management nearby, to blame for anything that goes wrong.

Eric Sprott had put a lot of money into a company called Pacton Gold, another Canadian junior lottery ticket. Greg Gibson was a director of both Novo and Pacton.

Quinton was working on picking up land in the central Pilbara region. The proposed acquisition was discussed with all the directors, and then to Quinton's great surprise, Pacton suddenly announced it was doing a deal on the same parcels. It could have been coincidence, and a stroke of pure luck for Pacton. Alternatively, it could have been double-dealing on the part of Greg Gibson. With my customary diplomacy, to both parties, I may say that the incident did not leave Quinton with enormous confidence in the business ethics of Greg Gibson. It looked as if someone had been telling tales out of school.

But behind the scenes Quinton had been doing some serious

wheeling and dealing for months, in order to pick up a couple of projects at Egina from local prospectors. (It is pronounced *Ed Ja Na*, by the way.) We had bought gold nuggets from Egina, from the metal detector dealer in Karratha. We compared them to the Purdy's Reward and Comet Well nuggets, and agreed they were virtually identical.

That was important. Egina was alluvial in nature; simply gravel, not hard rock. It would be easy and cheap to process, if you had water. Egina certainly had water during the monsoon season. It had so much water, you couldn't even move around the project. The other nine months it was as dry as a bone.

No one ever said Quinton's job was easy.

Novo was now sitting on the Beatons Creek property while advancing its far larger land position near Karratha. It was that phase in the lifespan of a junior resource company where progress is being made but, quite bluntly, it's dull progress. Gradually it was becoming obvious that there would be no more assays of sixty-seven grams of gold per ton gold, just a boatload of assays at two grams. Two-gram rock was a giant success because of the vast land position, but it paled in comparison to the early results.

On September 17, 2018 Novo announced two property transactions in the Egina region.[65] Only in hindsight would it become evident how important this was. Novo was paying $8 million in shares and cash to a company that held a mining license at Egina and had been mining gold nuggets from its alluvial gravels for years. Sincerely, I think I was the only person who really understood it. Clearly it conflicted with the direction that Kirkland Lake and Eric Sprott wanted Novo to take.

All investors want instant gratification, no matter if it is mom and pop with a hundred shares or a major with ten million shares. Novo set the bar so high with the first assays that everyone tended to ignore the considerable progress being made. Then, all of a sudden, it appeared that Quinton was off on a tangent with the Egina purchase.

Novo closed the deal two weeks later. [66] It would take a long time for the market to understand just how important the move into a large land position at Egina was.

Of course, this being Australia, the deposit there came with its own unique set of problems to be solved before any money could be made.

CHAPTER 20
KIRKLAND LAKE ATTEMPTS A COUP

I'M UNSURE HOW MANY VISITS I have made to Novo's Pilbara projects. My first was in June of 2009, even before its deal with Millennium Minerals on Beatons Creek. Since then I've been back almost every year until 2020, when the COVID-19 nonsense shut down the world's economy.

Since Quinton Hennigh is my best friend and we chat almost every week, I suspect I know more about Novo and about him than anyone else outside of the company. I also know more about Novo than many of the people in the company. And so I should, because this is a project I have been thinking about for a dozen years.

In late September of 2018 Eric Sprott, Tony Makuch, Greg Gibson and Quinton flew a plane Eric had chartered across the Pacific to Australia for an inspection trip on behalf of Kirkland Lake Gold (as one of the biggest shareholders). Eric is the Toronto billionaire who is also a big shareholder in Novo and probably the biggest shareholder in Kirkland.

They were going to take a look at Fosterville, and to determine where Novo stood and why not much seemed to be happening. Again, the initial assay results set the bar so high that almost nothing the company did would seem very impressive. To say that the atmosphere in the cabin was tense would be to understate it. If Quinton felt he was being ganged up on, he was probably about right.

Fosterville was looking more and more like a giant home run on Kirkland's part. Even though its merger with Newmarket had taken place less than two years before, Kirkland's share price had gone from about $5 to over $21 by this time.

Eric later spoke at a mining conference in Melbourne on October 3rd. Tony, Greg, and Quinton sat right at the front, in the first row. Eric talked about owning shares in Newmarket and Kirkland. He believed the two companies would become a powerhouse if they joined together. The market obviously agreed, with the shares up 300 percent. He talked about the discovery of the Swan zone at Fosterville, and how important it was in hindsight to the finances of the company.

I like Eric Sprott a lot. I have met him only once, and chatted for a couple of hours at best, but without question he has done more for Canadian mining than anyone I know over the last twenty years.

Because he's a billionaire, and they think differently, he made a giant mistake in his talk. He gave full credit to Quinton for the discovery of the Swan zone. Also for convincing Eric to proceed with the merger. He did this in front of hundreds of movers and shakers in the Australian mining sector in front of Tony Makuch and Greg Gibson. They hadn't made the discovery; they had thought it a waste of time and money and opposed the merger.

Tony Makuch and Greg Gibson demonstrated how no good deed goes unpunished. They were furious that Quinton got the credit for the work he did and they didn't do. They were determined to get even and if it destroyed Novo that was just tough shit.

Eric would make somewhere up to $1.5 billion out of the merger. He takes care of the people who take care of him. He told Greg to put Quinton on the payroll and pay him a nominal but regular salary. Greg did. It was six months after the merger between Newmarket and Kirkland Lake before Quinton was paid anything. Tony Makuch promised in writing to give Quinton options on sixty thousand shares, and in the end gave him fifty-three thousand.

After Kirkland Lake injected $56 million into Novo, Quinton realized he now had a conflict of interests. If he was president of Novo and was receiving a monthly salary from Kirkland Lake, just whom did he work for? He told Greg to stop the monthly stipend.

Quinton made a little money from his work for Eric on Fosterville, but given the ten billion dollar or so increase in the value of Kirkland Lake, it was pitiful. It was like tossing peanuts to the monkeys at the zoo. It was all because two functionaries were green with jealousy.

Quinton is the man in the arena, doing battle with everyone at the same time, including those who had made millions of dollars from his efforts and his skills. In a speech made at the Sorbonne in Paris in 1910, Teddy Roosevelt made an interesting observation about the man in the arena.

> It is not the critic who counts; not the man who points out how the strong man stumbles, or where the doer of deeds could have done them better. The credit belongs to the man who is actually in the arena, whose face is marred by dust and sweat and blood; who strives valiantly; who errs, who comes short again and again, because there is no effort without error and shortcoming; but who does actually strive to do the deeds; who knows great enthusiasms, the great devotions; who spends himself in a worthy cause; who at the best knows in the end the triumph of high achievement, and who at the worst, if he fails, at least fails while daring greatly, so that his place shall never be with those cold and timid souls who neither know victory nor defeat.

I was in Australia a month later, for another visit. While I liked the Beatons Creek story and was so-so about Karratha, I loved the Egina story from the beginning.

Beatons Creek was almost conventional. Karratha was anything but conventional, but it was obvious that it was a work in progress, albeit slow progress. Karratha's millions of ounces of gold will take from fifty to a hundred years to mine and process. The thousands of miners that all walked past the conglomerates over the course of a century had their very valid reasons for passing over the deposits.

Egina was perfectly conventional as a placer mine of sorts, other than the fact that there was always either too much water or too little. That's one of those niggling little facts that gets in the way of a good story. Not much was written about Egina because people simply didn't understand it. I went there in 2018 and saw the trenches that Carl Dorsch had put in, but beyond the fact that the pay streak was near surface, you couldn't tell much.

Carl had set up a simple plant near a hard rock shaft where gold had been produced decades before. There was some water to be had from the decline. He had holding ponds set up, but summers in Egina are hot. The ponds would quickly dry up until they could be replenished with the typhoon season's rains.

Quinton's sortie into the world of mechanical sorting would prove valuable. If sorting machines could separate gold from rock without the use of water, Egina could literally become a gold mine.

I've owned placer projects, and you can mine fairly low-grade material at a profit because you don't have to crush and pulverize it. Novo was on its way to picking up about 1,200 square kilometers of gravels. That's over 460 square miles. If all the gravel had gold in it, even at a low grade, there would be a lot of gold and a lot of profit.

Quinton and I visited Egina together. Carl Dorsch had left the fines from his small operation in 55-gallon drums. When we were bored we would pan the material. Unlike Karratha, there were a lot of fines. Our panning was always profitable.

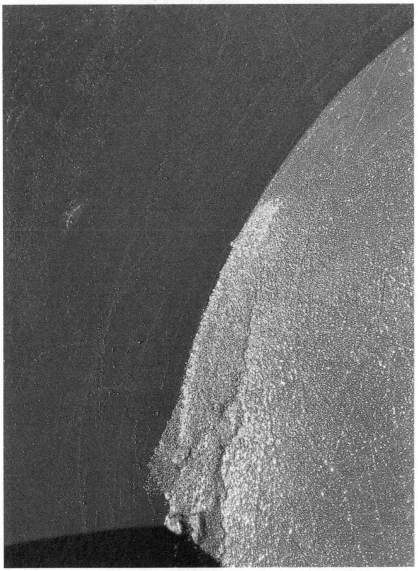

Fine gold panned at Egina, November 2018

We made a chopper tour of some ground belonging to Kairos Minerals. George Merhi showed us around the property. George, you will remember, was the chief geologist for Mark Creasy, with whom we had swilled $750-a-bottle wine on our tour of Beatons Creek back in 2009. George was now doing the exploration for Kairos, who had a project a little to the southeast of Novo's Egina tenements.

A few days before we walked the property, George and a mate had run metal detectors over an area of no more than two hundred square meters, and taken out over ten ounces of gold in three hours.

I cannot fathom why investors and commentators cannot come to grips with the remarkable potential of the Pilbara, which Quinton knew about twenty-five years ago. Where else on the surface of the Earth could two blokes make money so quickly? If any weekend warrior (I don't mean George) could do that with simple and relatively cheap equipment, what could a well-financed and well-managed mining company produce?

Quinton made some unsettling discoveries while we were at the Egina camp. Evidently, Greg Gibson had returned to Australia after the Diggers and Dealers gold show and, behind Quinton's back, was stirring up trouble. It appeared that he was telling people that Kirkland Lake would fire most of the Novo board and essentially would be running the company after its annual meeting, scheduled for the first week of December. Quinton was aware that Gibson was talking to Novo personnel, and it was probably not a good solution he had in mind for Novo.

John Youngson was running the exploration program at Egina and almost certainly would have been one of the people Gibson would be courting. Quinton confronted John, and I then understood what the "deer in the headlights look" actually meant. John protested and said that he was entirely innocent and would never stab Quinton in the back.

George Merhi's three-hour gold haul

I had met John and his partner Sue Attwood on the trip to the South Island in 2009. I had a lot of time for her. John's genius was mostly in his mind, although he had all the paper credentials of a PhD.

After my visit to New Zealand in 2009 I paid him and Sue to visit two of my placer projects, one in British Columbia and the other in Tanzania. I began to form the view that John lacked any feel for what he was doing. He could write a paper using all the proper words but he couldn't tell you where the gold was or how to recover it.

But I paid him a couple of hundred thousand dollars to source a placer plant and ship it to Tanzania for my operation there. The plant he provided would have been perfect for New Zealand, with its coarse gold, no clay, and lots of fresh water. Alas, I had fly-speck gold, lots of clay, and very little water. The plant was unsuitable. When I complained, John told me he hadn't designed it, just ordered it. But the purpose of having them visit the project before supplying the plant was to show them the conditions under which I had to operate. I wasn't a big fan.

When rot sets in, you must cut out that part of the structure. All Greg accomplished was to create ill-will among the people Quinton had spent years forming into a team. A number of people lost their jobs because they were confused about who was behind their paychecks. I felt bad for Sue as she had far more sense than to fall for a trick like Gibson's, but it appeared that John fell for it, hook, line and sinker. It cost him the greatest opportunity he would ever see.

After our quick trip to the Kairos ground with George, the chopper dropped us off at Port Hedland airport. Quinton and I returned to Perth. The next day, he and I and Rob Humphryson and Ronan Sabo-Walsh met in the Novo conference room to discuss what we thought was going on.

Quinton knew Tony Makuch and Gibson had it in for most of his board. Eric Sprott and Kirkland Lake had been busy in the area. They had put cash into Pacton Gold, of which Greg was a director. Pacton held ground in the Egina area that Novo had wanted to secure.

In addition, Kirkland Lake had put money into Artemis. As far as I was concerned, that proved that Makuch and Gibson were clueless. How dumb is it to deal with a company run by a pirate who makes it clear that he intends to stiff you at the first opportunity? The Boobsy twins weren't befriending the best and the brightest in Western Australia. They were climbing into bed with the gang of fools.

There had been a heated Novo directors' meeting in mid-November, and it seemed that Kirkland Lake wanted either to have Pacton take over Novo or to strip Novo's assets and hand them to Pacton.

These games go on all the time when a mid-tier company puts money into a junior. The tendency is always for the moneyed partner to believe itself to be smarter than the junior, and better able to make decisions. Almost always, that means better for the funding partner, not for the management or shareholders of the junior.

Quinton had been around long enough to have seen this game played before. Ever since starting Novo he had done his best to balance power between the potential vultures. And while Gibson and Makuch were quite impressed with themselves, I wasn't much impressed.

Eric Sprott had stakes in so many junior resource companies that he had to have someone representing his interests. When making cash injections he would often insist on having a seat on the board. That seat would be filled by Greg Gibson. This was a bad idea, for several reasons. Gibson was supposed to be running the Jerritt Canyon mine in Nevada but was also on something like twenty boards. I doubt he could have named half of them, much less provided management expertise. Eric's reach was exceeding his grasp.

When Eric came up with the idea of merging Kirkland Lake with Newmarket to achieve economies of scale, he needed someone senior to run the new, much larger Kirkland Lake. Greg Gibson nominated Tony Makuch, so there was always going to be a brotherhood of sorts between the two. Neither favored the merger, and made that clear. So there was conflict with Quinton, who fully approved of the proposed merger, foreseeing the new Kirkland Lake as a powerhouse. In the end he was proved correct, and neither Gibson nor Makuch has ever forgiven him.

I mentioned that tense flight from the U.S. to Australia. Progress appeared to be slow at Novo, and like all investors, Eric Sprott wanted the share price to be going up right now. Next, he was unhappy with Greg Gibson because Jerritt Canyon was turning into a money pit. And Makuch had been slow off the mark after taking over Newmarket.

The Fosterville mine was surrounded by a number of tenements important to the company. Australia has a loose way of controlling land. The properties were under the control of Newmarket but the expiry date on the tenements was approaching. Makuch should have made contact with the Department of Mines immediately and secured the land. He didn't, and the Department snatched the ground back and put it out for application.

When Rob, Quinton, Ronan and I met in Perth, another item for discussion was Novo's forthcoming annual meeting. Clearly something was going to happen, but what?

I am a neophyte when it comes to board meetings and annual meetings. I know they take place, but as with the manufacture of sausage, I just don't care to know the details. Ronan said something about shareholders abstaining. I had never heard the term before.

It seems that a shareholder can vote his shares either for a proposal or a director, or against, or can abstain. If more shareholders abstain on the matter of a director's election than vote in favor, it apparently means the director should resign.

When I found that out, I realized that even though Quinton and his people on the Novo board of directors could outvote Eric Sprott and Greg Gibson on issues because there were more of them, they could be booted from the board. In that event, Eric and Greg could then do as they wanted since they would become the majority.

I realized that Eric, Greg and Tony planned a backdoor coup. All they had to do was vote against the other Novo directors and the company was theirs to plunder.

Without a word to either Rob or Quinton, I returned to my hotel and wrote a fairly scathing piece about the Toronto Mafia. I suggested that all Novo shareholders abstain from voting on the matter of Greg Gibson as a director. I didn't suggest they did the same with Eric Sprott. I didn't want to attack him directly, only to fire a shot across his bow. I knew all the senior management at Novo and thought I understood their opinions of Novo's land position and Quinton's management skills. I thought Makuch and Gibson wanted to run everyone off the board except Quinton, Eric, and Gibson himself, thus giving all the power to Eric and ultimately to Pacton.

I pointed out in my piece that if the coup were to succeed and they canned Rob Humphryson, Quinton would leave as well. In fact the entire staff would walk out, leaving the keys to the office in an envelope taped to the door for the new management team. Jingle mail [67] works for companies as well as for homeowners. And just to let everyone know there was a price to be paid, I would lay the responsibility for the demise of Novo at Eric Sprott's feet.

When someone has a couple of billion dollars, money becomes almost meaningless. It's an accounting entry, not something real. How many houses, boats, airplanes and trophy wives can you keep track of, anyway? But believe it or not, billionaires are concerned about their legacies. If this triumvirate was determined to sink Novo, I wanted the world to know who to blame.

The vote took place in December of 2018. Makuch waited until the very last moment to vote the Kirkland Lake shares. He not only wanted to abstain on the three directors recruited by Quinton; he wanted Quinton gone as well. As I had recommended, there were many abstentions on Greg Gibson's election. Not enough to can him but enough to wake him up. Quinton twisted Eric's arm a little, pointing out that as a director he had at least a moral obligation to vote in line with management recommendations.

I am certain that my warning shot did remind Eric that while his vote could fire the entire board, the responsibility would be on his record. Discretion being the better part of valor, he voted along with the rest of the management team.

My piece may not have been the deciding factor in saving Novo but it did shed some light on a furtive attempted coup d'état. And I'm dead certain it made me many friends in the Toronto Mafia.

CHAPTER 21
TAKING ADVICE FROM KEYBOARD COMMANDOS

LEGEND HAS IT THAT STEVE JOBS of Apple Computer once said that the most wonderful thing about the Internet is that it gives everyone a voice. Alas, he continued, the Internet's biggest flaw is also that it gives everyone a voice.

Who knows if he did say it or not? He's not around to vouch for it but if he didn't say it, perhaps he should have.

We tend to forget that as little as twenty years ago there was not much to be found on the Web on junior resource companies. I began 321gold.com because I realized there was a demand for free information. The junior lottery ticket companies wanted their stories known, and investors wanted a choice. It was a perfect fit for us. Nova Gold (that's Nova, not Novo) was the biggest gainer on the Canadian stock market in 2001, and few realize the reason why. It was this: Nova was telling its story in a compelling way on the Web. I helped.

Doug Casey was still posting expensive paper newsletters with his stock recommendations to his subscribers. They got the information a week or so after we would have posted it on 321gold. Between the low in gold in August of 1999 and 2008, communication on the Web aided the creation of hundreds of new junior mining companies. They could raise money and attract investors because they now had a cheap, swift way to communicate. For years it seemed that every former taxi driver in Vancouver or drill crew supervisor was setting up his own company. Of course, given the 95 percent failure rate of mining projects, companies came and went regularly.

Over time, more and more investors would migrate to websites such as HotCopper and Stockhouse and CEO.CA. There they could read articles about new and upcoming mining companies and projects. In addition, many sites offered chatboards. There the punter was free to express his opinion, and all too often to attack anyone who disagreed with it.

Chatboards seem like a waste of time to me. I always wonder how many people would run their mouths the way they do if they didn't have an alias to hide behind. If they are so proud of their opinions, why won't they identify themselves?

When a guy has a keyboard and an opinion about something, and when the keyboard gives him access to a cast of perhaps thousands of readers, it seems only natural to start believing people actually care about your opinion. But keyboards and opinions are a dangerous mix. Having an opinion is not quite the same as having an opinion that anyone gives a shit about. Opinions are a lot like assholes. Everyone has one and most of them stink.

Quinton Hennigh thinks unconventionally. That is what makes him a genius. Unlike most geologists, he is not regurgitating the stuff that was fed to everyone in Geology 101. He thinks for himself. That's both rare and valuable. Miners and prospectors have crisscrossed the Pilbara for generations. No one put it all together until Quinton started Novo. Mark Creasy had a clue, and for most of the last thirty years was the biggest landholder in the basin. He didn't know exactly where the gold was but he pretty much knew there was a lot of it.

It should go without saying that the vast majority of people, and especially people possessed of a keyboard and an opinion, do not think outside the box. When they encounter someone who does, they have difficulty with it. Obviously, because they don't get it, then whatever is being done must be wrong.

Novo Resources is followed by a fair number of stock lovers who get it. It also has its full share of stock haters who clearly do not get it.

One of the most valuable posters in my view is on Stockhouse, and calls himself TX Rogers. His posts are thoughtful and intelligent. He understands the concept and largely agrees with the direction Quinton is taking. When he writes he adds insight to what can be fairly complex issues. He has no agenda, unlike other posters.

Another poster calls himself Rhino10. His real name is Ken Watson. He was a minor player in the Australian resource field and he hates Quinton. You see, when the managing director of Millennium Minerals made the deal with Novo on Beatons Creek, he didn't mention that he had also done a deal on the ground with Ken Watson and his partner for the alluvial rights. Since the reefs are basically unconsolidated gravel, there wasn't a world of difference between the rights Millennium sold to Novo and the deal Millennium did with Ken Watson and his partner.

For some reason Ken Watson and his partner broke up, leaving the alluvial rights to Beatons Creek in a sort of limbo. He blamed Quinton, and has spent the last nine years criticizing Quinton on chat boards, sometimes a dozen times a day saying the same thing again and again. But Quinton wasn't any part of the problems between Watson and his partner going their separate ways. Watson is so goofy that he came back in 2020 and wanted Novo to do a deal with him on some ground. That was pretty weird; you spend years badmouthing someone on a chat board and then ask them if they want to do a deal with you. I don't think that's going to happen.

It seems odd to me that Watson would have such hatred for Quinton and Novo. But he has posted hundreds, perhaps thousands of nasty grams on HotCopper and Stockhouse pissing on Novo's parade. And because he had sent me half a dozen meaningless emails ten years ago, he included me in his venom.

There should be some sort of check made, to prevent anyone from posting more than (say) five hundred emails all saying the same thing, to make certain the person posting isn't shithouse rat crazy. Watson wouldn't come close to passing. He is shithouse rat crazy. Saying so will probably set off a chain of another thousand stupid posts, all saying exactly the same thing.

Another very strange poster was someone named Taylor Dart who writes on occasion for Seeking Alpha as anyone can do. He wrote a pretty biased piece on Novo. That's not all that unusual. It happens all the time from those unable to think outside the box. But Taylor Dart then went to the CEO.CA chat board and proceeded to write one scathing post after another about what a lousy investment Novo was. That also happens. But he posted under a semi-alias of Trad. Someone noticed and pointed out that Trad is nothing more than Dart spelled backwards.

Trad had a fit and insisted he was not Dart. But if anyone posted anything aimed at Dart, Trad responded because he is Dart, just unwilling to admit it. How many times does a person have to lie before you realize he is a liar. Dart/Trad is a liar and that pretty much takes his credibility and flushes it down the toilet. He has an agenda and nothing more.

One change on the web over the last twenty years is the disappearance of paid subscription newsletter writers. There used to be half a dozen or so, and most have now retired or moved on to other ventures. Barb and I ran a computer website twenty-five years ago and we soon learned that the way to make something valuable on the Web was to give it away. Paid newsletter writers are caught in a bind. They pretty much have to tell people what they want to hear, or they will go elsewhere. Also, why should someone pay $500 a year for information available for free elsewhere?

One scribe based in South America had an interesting solution. Mark Turner is a self-hating Jew who supports terrorists in both Israel and the U.S. He instantly attacks anyone who even mentions the word Israel. He goes out of his way to make Jews look as evil and devious as possible. He whines and cries and pretends he is some sort of victim.

I caught him pumping and dumping a stock in March of 2009. I pointed out to him that pumping a stock under his own name for Hallgarten while trashing it on his other website was illegal. He lies all the time. He uses projection, accusing others of the self-dealing that he constantly engages in.

He is probably the worst financial forecaster I have ever read. Earlier in 2020, when silver dropped below $12 an ounce, he mocked everyone who bought or owned it. He continued his tirade as it rose all the way to nearly $30. He missed a 150 percent gain and was making fun of people on the other side of the trade the entire time. But he hates everyone, including the entire mining industry. People go to his admittedly meaningless website just to see Dr. Pimple Popper in action.

Mark Turner lives in Lima. I don't think he is there for the beaches. Lima has the ugliest beaches in the world. The nice beaches are on the other side of South America, in Rio. But Lima is a great place to hide out from the law.

Now I don't know if it is the same man, but the authorities in England are looking for a Mark Turner accused of child molesting. Almost certainly it is some other Mark Turner. And the authorities in Peru have a complaint of wife-beating from a former spouse of a Mark Turner. No doubt it is some other Mark Turner, that being one of the most common names in Peru.

Mark Turner would like to have an impact on the mining business. Perhaps he does have some influence. He makes the industry look as if it is run by and for evil people, with him being the leader of the pack. He is vicious, cowardly, and a stone cold liar. Everything he writes sounds as if it is coming from the mouth of either a drunk or someone on drugs. Clearly he has a chemical issue as well as being evil.

Naturally he is jealous as hell of Quinton Hennigh, who is everything Turner is not.

When I pointed out to him in 2009 that pumping and dumping, while momentarily profitable, was illegal, he threatened to sue me. I'm still waiting.

Mark: Please, please sue me. Because if you do, I will be able to take your deposition, and two minutes later it will be in the hands of the SEC and TSX, who will run your ass out of the industry.

CHAPTER 22
MOVING FORWARD TOWARDS PRODUCTION

THE ATTEMPTED COUP on the part of Greg Gibson and Tony Makuch did no permanent damage to Novo but did require some personnel changes. It didn't make sense to retain those who weren't entirely loyal.

Gibson's back door maneuvering left a bad taste in everyone's mouth and it was only a matter of time before he would be shown the door. Pacton Gold, the heir apparent, realized it now had no game plan. It began pulling up roots in WA and moved its operations and projects to Canada.

Kirkland Lake Gold achieved the highest grade of any gold mine in the world, without a shred of appreciation ever being shown to the guy that Eric Sprott credited with the success of the merger with Newmarket Gold and development of the Swan zone at Fosterville. It was as if there wasn't enough attention on offer for Tony to be able to share any with the man who handed him a gold mine on the proverbial silver platter.

Novo's stable of projects now had three distinct and different targets. Beatons Creek was near conventional. Since Quinton was waiting for Millennium Minerals to realize it had no way forward, it didn't make much sense to do anything other than publish an updated 43-101. That came out in April, showing just over 900,000 ounces of gold at an average grade of just over 2.6 grams per ton.

Novo had carried out test mining and had run material through a small plant, so was prepared if and when Millennium bit the dust. That 900,000 ounces was sufficient for five or six years' production, and in reality, how much sense does it make to spend your scarce money to define production material decades into the future?

The tough nut to crack was always going to be Karratha. Purdy's Reward and Comet Well were both permitted, but as long as David Lenigas was around, Novo wasn't going to advance Purdy's Reward an inch.

The problem remained how to define ounces to the satisfaction of the Department of Mines. Quinton made it clear in a press release on May 19, 2019.

> It is well understood that delivering a conventional mineral resource at the Karratha Gold Project is challenging owing to the extremely nuggety nature of the deposit. Accordingly, Novo has worked closely with independent experts (Mr Ian Glacken, Director of Geology at Optiro Ltd and sampling and geometall- urgical expert, Dr Simon Dominy) to ensure the Company's QA/QC processes and sample collection methodologies are sufficiently robust to underpin this mineralization report. Development of this minerali- zation report has become a guiding discipline to ensure Novo can pursue a mining lease application as well as update its NI 43-101 technical report for the Karratha Gold Project.

Quinton was juggling two balls in the air. The more difficult was determining how to measure the unmeasurable. Novo had already extended the length of the conglomerate reef for over ten kilometers, including both Purdy's Reward and Comet Well. Once the Department was on side with the process — drilling for the structure of the conglomerate reefs, and taking bulk samples to determine grade and distribution — Novo would hopefully qualify for a mining lease.

Quinton then needed some cheap, swift method of mining and processing the hard rock.

Ever since the first samples were taken at Purdy's Reward, some two years earlier, Quinton had been working with Steinert, a German company, on the use of its mechanical sorters.

The Steinert XSS T mechanical sorting machine used X-rays to identify the particularly dense gold in rocks. When it encountered a rock containing gold, a puff of air would pop it into the ore bin for later processing. The process was highly successful but Novo spent a lot of time deciding just how to crush the particularly hard rock conglomerate, and to what size.

One of the original samples, taken in 2018, was for some reason crushed to 2 mm size and sent for processing. That is a pretty small rock, and it overloaded the capacity of the machine to sort. By the time someone realized the error, tons of bulk sample material had already been crushed to 2 mm. It's easy to make mistakes when you are working on something entirely new, attempting something that has never been done successfully before. But investors are impatient with any error, and raked both Quinton and Novo over the coals.

Initial tests of the Steinert machine showed it could concentrate the rock down to less than two percent of its original mass. That was a game changer for Novo and for the entire mining industry, as it meant that projects previously assessed as uneconomic could be economic now. Much of the cost of mining is incurred in moving around tons of material, again and again. If the crushing and sorting could be carried out near the mine, and if only two percent of the mass need be moved around, costs would be far lower.

But even the Steinert machine had limits, and it didn't like all the material being of 2 mm size. Rocks that small would blank the sensors. If it worked at all, it was like molasses in winter.

Later in 2018, Novo began to test mechanical sorting machines made by Tomra, a company formed in Norway decades ago but now based in Germany. Novo had realized that the nugget effect caused a giant problem with the accuracy of assays, since the nuggets were distributed at random, both horizontally and vertically. However, that same issue was also an opportunity, if they could pick the correct machine for mechanical sorting. Nuggets were hard to measure but easy to process.

The Tomra machine used a combination of X-ray transmission (XRT) and electromagnetic (EM) induction. The former could detect high-density material while the latter was in effect a metal detector. Gold nuggets have a specific gravity of about 19, depending on what other minerals they contain, so were perfect for both XRT and EM induction.

Novo had been testing the Steinert machine since work began at Purdy's Reward in 2017. They were quite satisfied with the 98 percent reduction in mass, but with the Tomra device, even the initial tests on material of 10–63 mm size showed a 99.7 percent reduction in mass. From a technical point of view, that was a home run. For every 2,214 pounds of input (one tonne), the sorted gold was about 6.6 pounds, or three kilos.

Novo continued to test both machines until well into 2020, with wonderful progress. The vendors of both the Steinert and Tomra devices worked closely with Novo. As experience grew, both showed constant improvement in the potential of their machines, both in capturing as much of the gold as possible and in reducing the sorted mass to the lowest fraction of the original amount.

The gold from the Farno deposit at Egina almost demanded the use of mechanical sorting, if the necessary economies of scale were to be achieved. Egina and Karratha material of various sizes were sent to both Steinert and Tomra for testing. They soon found that rocks larger than 63 mm were simply too large for the machines, while rocks smaller than 6–10mm were too small for the machines to work well.

The new projects in the Egina area were being largely ignored by Novo investors. They were different, and as difficult in their own ways as the Karratha story.

Quinton had done a deal to acquire the Farno McMahon tenements as early as September of 2018. He had been working behind the scenes for a year to pick up the tenements, realizing that the size and shape of the gold nuggets were remarkably similar, from Purdy's Reward to Egina. But Karratha was hard rock and Egina was fairly loose gravel.

Carl Dorsch had been mining at Egina and selling the gold to the metal detector guy in Karratha. That is where Quinton and I bought gold in mid-2017.

Carl was doing the most primitive mining possible. He stripped off half a meter to a meter of sand and other material at the surface, then worked in a layer of gravel sediments 1–2 meters thick that contained the gold. Using a metal detector, he would have someone mark the hot spots and they would dig up the nuggets. It was a cheap method, and more or less effective.

Quinton needed to find a way to test various areas to see if they were prospective for gold nuggets, and find a low cost method of processing the gold-bearing gravel.

Again — and you will hear me say this a lot — Quinton's ability and willingness to think unconventionally meant that it might take some time to identify a good way forward, but it would work and it would be profitable. Naturally, most of the people in the industry would mock him, but that was because they weren't smart enough to figure out the problems, analyze them, and devise a practical solution.

Carl Dorsch had been mining at Egina for years

What is most remarkable about the Pilbara region is the long history of gold mining but the limited production of gold. Many people had worked the Pilbara; they just hadn't thought outside the box.

Carl had a small fleet of trucks to carry the unconsolidated gravel to a small plant he had back at his camp, for further processing. It was not efficient, as most of the gold had been located and removed as nuggets. But if you looked at his costs and the amount of gold produced, it was easy to see how much more efficient the process could be made to be.

The camp was built at the site of a former underground gold mine. The adit contained water after the monsoon season. As long as the water lasted, Carl would process material.

Quinton sought a better way.

The first thing he had to do was to get Heritage Clearance from the natives on the areas outside the Farno tenements. That has become a slow process. Prior operators had reached agreements with the two local native groups who controlled the region, but then failed to do what they said they would do. As a result, the locals weren't very trusting or easy to work with.

The process could take years. Eventually the natives will realize the financial potential. There is a lot of gravel-bearing ground, from Egina to the Indian Ocean. A fair and reasonable royalty would change their lives for the better, for a very long time. It hadn't happened in 135 years, but with Novo and Quinton Hennigh it would.

Without gaining Heritage Clearance, Novo could not take samples or disturb the ground. But it could test the two pieces of tenement that had been granted mining rights, and interpolate those results to the rest of its Egina properties.

One instant solution that did not require a Heritage Agreement to be in place was the use of ground-penetrating radar, to determine the very slight differences between the half-meter of overburden, the gold-bearing reef of 1–2 meters of loosely consolidated gravel, and bedrock. The radar didn't disturb the ground at all. It made quick work of determining which areas should be the most favorable for mining.

The next step was to determine grade. In 2019 I visited Egina for the second time. Keith Barron was on the same tour. We also brought Erik Wetterling with us, a young Swedish writer who had been covering Novo for a couple of years.

Keith has a sapphire mine in Montana so he knows full well what it costs to process a cubic meter of alluvial material. I've had several placer mines from British Columbia to Ghana, Tanzania, and Sonora, Mexico.

When you talk about the costs of mining alluvial material, it is industry practice to use cubic meters rather than tons. The yellow gear that is used to move the gravel around is all denominated in cubic meters — the capacity of a shovel or bucket or dump truck. The weight of the material is meaningless. Keith and I sorta agreed that costs would be in the range of $10 to perhaps $15. But when we later saw the progress made with the latest and greatest sorting machine, in Perth, and then went on to Egina and saw how close to the surface the gold-bearing gravel was, we felt that $5 to $6 U.S. per cubic meter was a good target number for Egina.

On December 20, 2018 Novo announced mechanical sorting results from the Tomra machine, using ore from Comet Well. The concentrations were as little as three percent of the original mass. That's the two kilograms per ton result from rocks in the 6–63 mm size range.

In addition, but missed by almost all investors, were the alluvial tests from the Egina gravels showing 107.88 grams of raw gold from 95 cubic meters of gravel. Guessing a purity of about 93 percent fine gold, that works out to 1.06 grams per cubic meter.

At the time a gram of gold was worth about $50. So you have a basic mining and processing cost of perhaps $5 to $15 a cubic meter to make $50.

When Quinton and I visited the South Island of New Zealand in 2009 we looked at projects where owners were delighted to mine 0.3 grams per cubic meter. There were areas at Egina that were returning over three times as much. One gram per cubic meter would be an excellent result for any alluvial mine.

Quinton was thrilled to get numbers as high as his team achieved at Egina. They used the ground-penetrating radar to find the swales where the highest-grade material should be. I was of a mind that they really didn't need to do that. My preference would be to test what should be the lowest-grade gravels. We knew that there was gravel pretty much from Nullagine to the Indian Ocean. From Egina, a straight line to the ocean at its closest point was seventy kilometers long. It was virtually all gravel and mostly flat, with a few hills with conglomerate reefs outcropping.

Novo's outright land position was of about 1,200 square kilometers, or over 460 square miles. Adding the ground belonging to De Grey Mining, on which Novo had an option, took the total to more like 2,500 square kilometers or nearly 1,000 square miles.

If a square kilometer contained a one-meter layer of gravel bearing 0.3 grams of gold per cubic meter, that would equate to over 9,000 ounces of gold per square kilometer.

Of course, gravel layers will not be found beneath the entire 2,500 square kilometers, and some of the land is hilly. Neither is there any evidence that all the gravel will be as high-grade as 0.3 g/m^3, but my guess is that a lot of it would be. Obviously the +1 gram material will make the most money, but even 0.3-gram material would make a 50–66 percent margin. Every mine manager in the world would love that.

All investors should be required to memorize the Lassonde Curve before being allowed to make their first stake in a junior resource company. All stocks go up and all stocks go down, but resource stocks tend to follow a fairly predictable pattern that the remarkable Pierre Lassonde first identified thirty years ago.

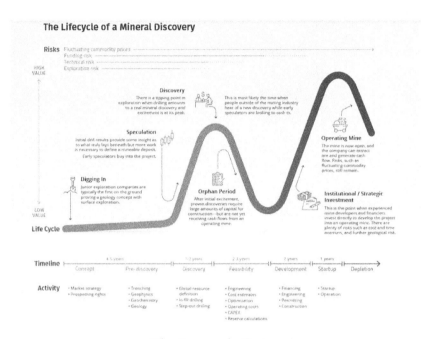

The Lassonde Curve

It took from 2009 until 2017 for Novo to get traction with the discovery of conglomerate reefs bearing gold nuggets in the Karratha region of Western Australia. After its share price had meandered between around fifty cents and $2 between 2011 and 2017, Novo shareholders hit the jackpot in 2017 when the first assays were released from Purdy's Reward. The shares peaked at $8.83 in late 2017 before tumbling by as much as 85 percent into the crash of March 2020, when investors finally understood that the world really was coming to an end.

But it didn't, and Novo has since been in the boring period of the cycle, after a peak powered by speculation and a later realization that maybe it wouldn't continue to rocket higher. All juniors do the same thing. From 2017 until well into 2020 the stock price climbed at times and declined at times, just like every other stock.

From 2018 into 2020 there was progress in defining grade and tonnage at Beatons Creek and Egina. Constant progress with the mechanical sorting machines from both Steinert and Tomra improved their ability to pick up smaller and smaller gold, along with a most dramatic reduction in total mass. At least for Purdy's Reward and Comet Well and the surrounding ground, profitable mining was possible only if the gold could be removed from the dross at an early stage. Finding it would remain an issue, but clearly, the use of mechanical sorting machines made profitable mining possible there for the first time.

We reached the first peak in the Lassonde Curve and are now in the "orphan period" prior to the start of production. While many investors moan and groan constantly about the lack of important news, and other investors with an agenda will constantly bash Novo, actually Novo offers a gift to investors every year. That's equally true for the bulls and the bears.

I have never noticed any other newsletter writer or commentator mention this, but the junior resource market gives small investors an edge that exists in no other market. Small investors in resources have a great advantage over the large funds and the deep-pocketed investors. At the worst of times and the best of times, the tiny traders can get an execution. Now it's true that at market tops there is total liquidity and at bottoms it's hard to get a fill. But it's still possible to buy or sell $5,000 worth of any stock at any time. The only time a large investor can get a fill on, say, a one million share block of anything is at market tops.

The juniors have a trading range that almost all investors ignore. It's something I discuss in my books. It is common for the price of a junior stock to cover a 300–1,000 percent range in a single year. (I mean the difference between its lowest and highest prices in the course of a calendar year.)

Novo has a smaller range, about 300 percent usually, but that means that if an investor were to buy Novo near a yearly low and was willing to hold for a 50 percent gain, it would happen every year. Likewise, if an investor went short near a yearly top in the shares, he could make a 50 percent profit every year. Where the vast majority of the investors I am aware of go wrong is to hold for a 1,000 percent gain.

It's true that they happen. I said in 2012 that Novo would be a 10–100 bagger. So far it has gone up only 3,600 percent. But anyone, bear or bull, could have made 50 percent on Novo every year, and those wins add up over time.

The real advantage of Novo over other juniors is that it has great liquidity. Many juniors trade only by appointment at lows. You can still make the same profitable trade. I do it by putting in stink bids at ten percent under the lowest price of the day before. No stock really loses ten percent of its value in a day without announcing terrible news, but at lows the weak hands love to dump the shares they bought at the highs.

Novo is in that part of the Lassonde Curve where serious investors make the most money with the least risk. It would be worth printing any of the variations of the curve you can find with Google, and keeping it handy.

In early 2019 Novo made an interesting announcement,[70] saying that Sumitomo Corporation had agreed to provide up to five billion yen (about $46 million at that time) on a mutually agreeable project, and as a result would have a right of first refusal (ROFR) once it had made an investment, should a third-party investor make an offer for all or part.

While it was true that the press release didn't actually commit either side to a particular deal, it did indicate an interest, on the part of one of the foremost mining and metals companies in the world, to be a part of something that Novo had. The Japanese tend to take a long-term approach to dealing with other companies. They do not invest in the flavor of the month, and much internal discussion and research will precede an approach to another company. It was significant, for several reasons that will take years to fully understand.

In April of 2019 Novo announced a 30 percent increase in the resource at Beatons Creek, including just over 750,000 ounces in oxide and fresh rock mineralization.[71] The oxide portion, of 325,000 ounces, was especially suitable for the Millennium Minerals mill, when that might become available.

May of 2019 brought the ultra-important environmental approvals to mine the Beatons Creek deposit.[72] In addition, all Native Title and tenure documents were in place in anticipation of mining. These agreements and permissions are what often delay junior mining companies in moving forward. Novo was ready to mine. All that was needed now was a mill.

June of 2019 saw an agreement signed whereby Sumitomo would spend up to $30 million U.S. over three years at Egina to earn a 40 percent interest in Novo's project there, including the Farno tenements and the ground included in the joint venture between Pioneer Resources and the De Grey Mining.[73]

Late in June came a binding letter of intent with De Grey, with Novo to explore the lag gravel deposits on ground held by De Grey. De Grey concentrates on hard rock underground deposits in the region, so was happy for Novo to pick up rights to the 1,100 square kilometers of gold-bearing lag gravels.

In August of 2019 a Novo press release announced just over one gram of gold per cubic meter of gravel at Egina, from four bulk samples totaling 282 cubic meters.[74] A purity of 89–95 percent would equate to 0.9–0.95 grams of pure gold per cubic meter; excellent numbers, and potentially very profitable.

Novo took a batch of nuggets from Egina and sent them to Germany, for Steinert to test on a machine using a new technology called an eddy current separator. The tests were announced in a press release in September of 2019.[75] Novo fully understands that if it is to succeed in mass production at Egina, it must use some form of dry processing with a mechanical sorting machine. Novo has worked closely with Steinert for over three years now.

October of 2019 saw more bulk sample results from Egina: 222 cubic meters of gravel produced 337 grams of gold nuggets when processed, or just over 1.5 grams per cubic meter.[76] Guessing a purity of 90 percent would mean a solid 1.35 grams of pure gold per cubic meter, worth about $67 U.S. and costing perhaps $10 to recover, even if Novo does a really rotten job of mining and processing.

Investors continued to ignore these incredible results. That would be a world-class number in any placer operation anywhere. Given that the Egina area demands dry processing, Quinton has done a series of deals on over 1,100 square kilometers of gravel-bearing ground with exceptional results. My view is that Egina will in time become the jewelry box for Novo.

Keith Barron and Erik Wetterling had visited Novo at its Karratha projects, but neither had been to Egina or to Nullagine. We joined Rob Humphryson for a short tour of all three of Novo's major gold deposits in October of 2019.

By this time Karratha was on a sort of hold, since a fair bit of money would be needed to advance it further, we knew where the gold was. We knew how to separate it from the country rock. Novo needed a mining permit, and that was in the works. But any further progress was going to eat money.

Quinton was at home in Colorado, awaiting the birth of his second grandchild. He had strict instructions to stay there until she arrived.

Keith Barron and Rob Humphryson at Egina, 2019

I looked forward to this trip with my friends, partly because I wanted to take the measure of Rob Humphryson. We were close to the point at which his work would become more important. Wherever in the world a mineral discovery is made, someone has to put it into production. Quinton had done 90 percent of the hard geological thinking, but someone now had to put the pedal to the metal to advance matters from theory to practice.

You have to have a few beers with someone over a period of a week or so and share some lies before you start to understand what they are and are not capable of. Rob was a great guy to go on a tour with. He told lies with the best of us. If he would just learn to hold his beer a little better he could be a good companion on a site visit.

Erik Wetterling had begun following Novo a couple of years before. On his own, with no prompting, he had written a series of short pieces talking about the company, and why he saw it as a great investment.

I had previously suggested to Quinton that he bring Erik over to Western Australia. Jay Taylor and I were about the only other writers covering Novo. Brent Cook pretty much wrote off Novo after the disappointing assay results in late 2017, but he had never visited Egina. Quinton hadn't picked up any of that ground when Brent washed his hands of Karratha.

Erik brought a new pair of eyes to the story, and a different way of looking at the company.

Novo brought its IGR3000 test gravity plant from Beatons Creek to Egina and fastened it to the plant operated by Carl Dorsch. It worked well, and recovered even the tiny gold not picked up by the original plant. The gold from Egina was quite remarkable, and at times Novo recovered nuggets larger than one hundred grams.

Erik Wetterling made his first visit to Egina

One thing I want readers to think about is the soil attached to the nuggets, and its solid red color. That is all iron staining. Later in the book I will explain why, if you understand how the iron got there, you will also understand at once the giant scale of the gold deposit in the Pilbara Basin.

Egina test plant with IGR300 for fine gold

Since most of the gold at Egina was in the form of nuggets of decent size, either the Steinert or the Tomra sorting machine would pick up all of it. Tests had also shown that the machines would even pick up gold of near pinhead size.

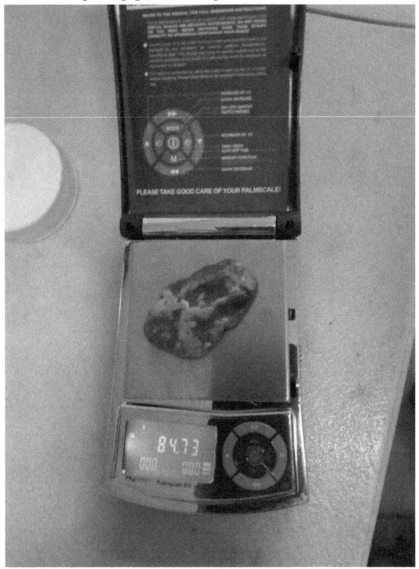

A not uncommon large gold nugget from Egina

Of the three projects, Egina impressed both Erik and Keith the most. Without seeing how Novo prepared the ground for mining, and without seeing an actual cut showing the nugget-bearing layer of gravel, it wouldn't mean much. Since Keith mines alluvial sapphires, he realized how inexpensive the mining could be.

Typical nuggets found with a metal detector

First, you strip off the thin layer of overburden. Novo intends to literally slide a steel plate under a section of overburden and move it to the side. Then, process the meter or two of gold-bearing gravel (marked "GV" in the photo). The loose gravel is run through either a Steinert or a Tomra sorter and put right back where it just came from, minus the gold picked up by the machine, which is fed into a container of its own, never to be touched by human hands.

At the end of the process the overburden and vegetation is put back in place on top of the gravel. It will cost more in operating expenses to reinstate the ground than it will cost to mine and process the gold.

There is a clear difference between the gold-bearing gravel and the basement rock. While the reef does pinch and swell, some small amount of basement will be processed. Nuggets can be found into a few centimeters of the basement.

Rob does enjoy showing visiting firemen around Egina, since it cannot fail to impress. The exploration crew were out with metal detectors, scanning every level of the reef and carefully measuring what they found for their records. Rob had us give the metal detector a go. We each found a nugget. We could hardy fail to, there being so many of them.

While the grade wasn't as good as in the reefs at Karratha, it was remarkable to me how similar the stripped ground looked when the nuggets had been mapped. Except that Karratha was hard rock and Egina was unconsolidated gravel, easy to process.

Overburden has been stripped; "GV" is gravel with gold

Karratha and its many kilometers of outcropping conglomerate reefs will contain far more gold than in the area from Egina to the Indian Ocean, but Egina will be the lowest-cost gold mining operation in the world, and the easiest to process.

Keith Barron was quite comfortable with the estimate of $5 to $6 a cubic meter. Most of that cost will be in stripping and after-mining remediation. The worst case looks something like $500 to produce one ounce of gold.

Bob Moriarty metal detecting at Egina

I am exceptionally willing to suggest that Quinton and Rob have opened a whole new world of low-cost mining though the use of effective and cheap mechanical sorting machines. For that alone they both deserve the thanks of the mining industry.

After we had visited all three projects with Rob, we returned to Perth for a demonstration of the Steinert machine in operation. It was remarkable. It could pick up pinhead-size gold. Since the tiny gold is only a small fraction of the Karratha conglomerate story, that will not be a significant contributor to project economics, but in the Beatons Creek conglomerates there is a lot of small gold.

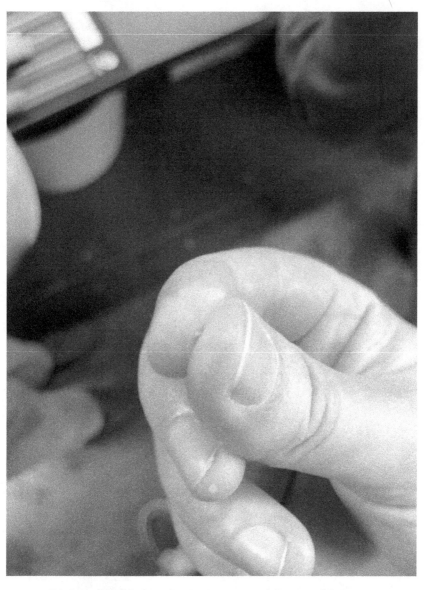
Pinhead gold that the Steinert machine could detect

Steinert machine in Perth, October 2019

We, all three of us, found the technical capability of the machine quite incredible. But when you delved into the operating costs it became even more interesting. The biggest expense was operating the air compressor that popped the gold into the ore bin. (A machine costs about $750,000 in Australian money.)

The rock, crushed to the correct size, ran up a conveyor belt, over the rotating sensor wheel and into the sorted material bin — except for the gold, which was popped right into the gold bin. It would be untouched until the sealed and weighed lockbox was delivered to the refinery. The cost per ton of sorted material was going to be tiny.

CHAPTER 23
MILLENNIUM GOES TEATS UP

THE NEWS THAT QUINTON had been waiting patiently for arrived a month early for Christmas in 2019. It was a wonderful present nevertheless. He had known for years that it was coming, but the price of gold would inch higher and give Millennium just a bit more breathing room. Finally it arrived. [77]

Millennium Minerals moves into voluntary administration
© November 25, 2019   ■ News   ▲ Vanessa Zhou

This meant that Novo had to shift direction at once. Work would continue at Egina in determining the grade and extent of the gold-bearing gravel. Karratha would remain on the back burner, and Novo would move towards the purchase of Millennium's mill and going into production at Beatons Creek.

Millennium's bankruptcy was interesting because of the legal structure. IMC, a Singaporean company, had advanced about $69 million to Millennium, using the mill as security. When Millennium went into bankruptcy IMC took over the company.

Novo had to come to an agreement with IMC if it was to take over the mill. But IMC wanted to sell the entire company, which had possible advantages for Novo as there were numerous land packages included.

Quinton spent from November of 2019 until August 4, 2020 finalizing an agreement. The mill had cost about $100 million to build but had of course been used, and poorly maintained, with the management of Millennium just trying to keep the wolf from the door. In a way it was a Mexican standoff, where two knife fighters tie their left wrists together before starting to battle. If Novo didn't buy the mill, IMC would have a pile of rust to get rid of, but Novo would have no quick route to production.

IMC actually made out like a bandit but could have done far better still. They should have taken the proceeds in Novo stock but chose to take $60 million Australian in cash. I'm always baffled by the actions of those running big companies. It seems to me that the bigger the company, the dumber the decisions. Millennium Minerals had never been anything more than a supplier of jobs to mining and mill personnel. Why would IMC be happy to keep throwing money at Millennium, but prefer cash to Novo shares?

Now Quinton could have the production he had been trying to get for so long. Building a mill would have taken five years to permit and construct and would have cost well over $100 million.

Running a production company is a whole different kettle of fish than operating a junior resource exploration company. As an exploration company you can never fail. You can always whine that all you need is a few more bucks for another drill program which is sure to hit this time. As a production company, it's either deliver or get fired. It's a lot like the difference between ham and eggs. Now the chicken, she's involved. But the pig, he's committed.

Quinton had to consolidate the company and make whatever changes he had to make before production started. The COVID-19 virus that arrived early in 2020 actually helped Novo. Rob Humphryson, an Australian in Australia, could oversee the refurbishment of the plant and the hiring of people and assembling of equipment. Since Quinton and everyone outside Australia were essentially locked out of the country for the best part of a year, Rob could do the job he had been hired to do without someone looking over his shoulder and offering advice. It was his baby to screw up, or he could look like a hero. I have a lot of confidence in him and believe that hiring him was one of the best decisions Quinton ever made. But if he screws this up, I'd fire his ass.

So while Quinton was operating behind the curtains, working with IMC on a plan and a price for the mill, he started spreading his wings in other directions.

Karratha was under control, but would remain on the back burner until Novo had the cash flow generated by going into production at Nullagine with the Beatons Creek material. At Egina, various areas were being tested with the ground-penetrating radar. Research continued with the Steinert and Tomra sorting machines.

As for the Heritage Agreements for the other ground, Novo has still to reach agreement with the two native corporations involved. Since the advent of COVID-19 they have hunkered down, and little progress will be made until the parties get together.

But Quinton always wanted to have at least a mid-tier gold mining company, and maybe even a major. In early January of 2020 Novo announced the purchase of ten million shares of Australian-listed Kalamazoo Resources.[78] Eric Sprott took a similarly-sized chunk of shares. The private placement was done at forty cents a share, with a full warrant at eighty cents for eighteen months.

As I write, the shares are up about 50 percent, but hitting the warrant price in another six months might be a bit of a struggle. Kalamazoo looks as if it has its hands on another Fosterville. Time will tell.

February 2020 brought yet another brilliant bulk sample from Egina in the gold-rich swales.[79] This one showed 562.25 grams of gold and nuggets from a 413.6 cubic meters. Alas, only Novo, Keith Barron, and I understand how rich a test result of 1.36 grams per cubic meter is. Figure maybe 1.25 g/m^3 of pure gold, and today's gold price of $60 U.S. a gram, with costs of $5 or $6 a cubic meter for a 90 percent margin.

Getting Egina into production is a priority, but Novo must work within the confines of what the natives will allow. One day soon we can hope they will realize it would be nice to have steady dividend checks coming in monthly to each corporation.

The results of testing the Tomra machine in Sydney with Egina material showed 100 percent recovery of the gold in rocks in the 6–18 mm range, and 100 percent recovery from rocks of 18–50mm size.[80] The Tomra sorter really likes the 6–18 mm material, with 100 percent recovery while rejecting 99.7 percent of the waste rock. They could run the 6–18mm material at a rate of about ten cubic meters an hour, and the 18–50mm material at about twenty cubic meters an hour.

In March of 2020 Novo made a major investment in a tiny junior in Newfoundland, trading 6.94 million of its shares for fifteen million shares of New Found Gold.[81] New Found later went public, and as I write in January of 2021, those same Novo shares are now worth $16.53 million, while the shares they picked up in New Found are worth an incredible $61 million.

It took Artemis only two and a half years to wake up to the fact that by stiffing Novo in 2017 and demanding $20 million in hostage money for the "binding" letter agreement signed by David Lenigas for Purdy's Reward, Artemis had only ensured that Novo wouldn't drop a dime into the property as long as Artemis held it.

In March of 2020 Artemis announced it would sell its half of Purdy's Reward to Novo, as well as the 47K property, for two million Novo shares and $1 million Australian. The Artemis–Novo joint venture was now a thing of the past. Novo would now control 100 percent of Purdy's Reward.[82] David Lenigas was long gone from Artemis, having sold all his shares at the top, but his legacy still left a stench.

In late March of 2020 Novo announced it had taken a stake of nine million shares in Australian-listed GBM Resources, with an option to earn up to a 60 percent interest in GBM's Malmsbury project, located in the famous Bendigo gold belt in Victoria province.[83] Novo would earn in with a combination of shares and gold exploration expenses.

With the acquisition of the Millennium mill under discussion, Quinton wanted to consolidate as much potential ground in the Nullagine area as possible. In June Novo announced it would acquire nineteen square kilometers of ground in three exploration licenses from the vendors of what was called the Mt. Elsie project, about 75 kilometers from Nullagine.[84] It cost Novo 324,506 shares and $100,000 in cash. Novo now controlled a total of about 13,750 square kilometers of ground in the Pilbara Basin. That's over 5,300 square miles; nearly one Connecticut or two Delawares, but hopefully containing a lot more gold than either of those states.

Novo's technical team at Egina began to use a new device in 2020, called a mobile alluvial Knudson or MAK.

The difficulty with such a giant land position is how to home in on the highest grade, or the material that is easiest to test, rather than just using Kentucky Windage to guess where the gold might be. The MAK unit takes grab samples of approximately one ton or 0.4 cubic meters of gravel at regular intervals from a pit 1–2 meters deep, and counts the gold grains. In July of 2020 Novo announced it had taken 342 such samples, with another 750 scheduled. [85]

The purpose of the MAK sampling is to get a feel for the gold in a certain area, and then follow up with 120 bulk samples of 30–50 cubic meters of gravel for detailed processing through the IGR3000.

Due to the nugget effect, the MAK samples will provide only a rough idea of grade. The use of the IGR3000 will tend to correlate where the MAK is most effective after the larger bulk samples are processed through the plant at the camp. The MAK processing is done in the field. The IGR3000 bulk sample gives an exact amount of gold in a bulk sample, but takes a lot longer to schedule and to make arrangements to scoop out the gravel and deliver it for processing.

By far the most important press release in Novo's history came out on August 4, 2020. [86] It announced the agreement to acquire the mill of the defunct Millennium Minerals and its 230-man camp for $44 million in Novo units priced at $3.25 a share and a half warrant at $4.40 good for three years. Following Novo's acquisition of the Millennium shares, Millennium would repay IMC $43.3 million in cash and 6.5 million Novo units. Novo was providing Millennium the difference between what Novo is paying for the shares and what Millennium must repay to IMC for its secured debt.

Novo needed cash for the transaction, and so raised a total of $56 million in private placements and secured up to $60 million in debt financing with Sprott Lending, in two parts. The sum of $35 million was to be advanced at closing, with an additional $25 million available up until March 31, 2021 contingent on Novo completing a pre-feasibility study on Beatons Creek that is acceptable to Sprott. The debt bears an eight percent coupon plus LIBOR or one percent, whichever is greater. Repayment begins twenty-four months after closing, in equal quarterly installments.

To put this in terms investors can understand, Novo is on the hook to Millennium and IMC for about $65 million, plus whatever it takes to put the mill back into operating condition. It is also on the hook to Sprott Lending for an additional $35 million.

The mill has averaged 1.88 million tonnes of throughput per year for the past five years. I've made it clear that I didn't think much of Millennium's management. I do have a high estimation of Rob Humphryson and his crew. If all Novo does is to equal Millennium's throughput, feeding in rock averaging two grams of gold per ton, that will result in the production of about 120,000 ounces of gold a year. I should be sorely perturbed if Rob cannot beat that, and in such an event I will not buy him any more beer. I think the all-in cost of production will be in the $700 to $900 U.S. range. But what do I know?

In October of 2020 Novo announced progress in Perth, with the use of a Steinert KSS sorting machine for use in bulk samples taken from both Purdy's Reward and Comet Well. [87] This would start in March or April 2021, after the worst of the cyclone season. The machine had been ordered and was due to arrive in Perth in November. It would then be necessary for Steinert personnel to assemble and finish the machine according to Novo specifications before it is taken to Karratha and put into use.

As of early January 2021, as I write, Novo is still on track to do the first pour in about mid-February. I plan to release this book on the day of the first pour.

Investors wondering about what to do next might want to look again at the Lassonde Curve for a hint.

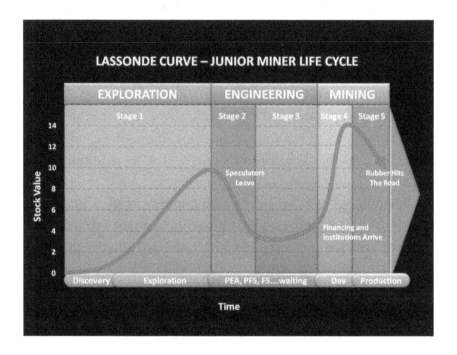

CHAPTER 24
IT'S NOT ROCKET SCIENCE

NO DOUBT THE NAYSAYERS will be out in force, telling everyone just how stupid Bob Moriarty was for calling this the greatest gold discovery in history. But then, unlike me, they weren't saying that Novo was going to go up 10-fold or 100-fold when its shares were forty-five cents, so maybe I'm not quite as stupid as they will say. I did get old but I didn't get stupid.

You see, I am unencumbered by what some geology professor told me in Geology 101. I never took Geology 101, or any other chemistry or mining course. I had to learn to think and to see what was right in front of me. If you can develop that ability you will see the same thing.

I was actually part of the Novo story as early as 1976, long before Quinton came up with his theory. I didn't know it at the time; I thought I was just a ferry pilot (do get the spelling of that word correct) taking a Rockwell 685 from Southern California to Melbourne for Lang Hancock. [88]

Hancock was the guy who jumpstarted the entire Western Australia iron boom. While he claimed to have discovered the world's largest iron deposit, it was actually Harry Page Woodward who did so. [89] In 1890, in the *Annual General Report of the Government Geologist*, he wrote:

> "This is essentially an iron ore country. There is enough iron ore to supply the whole world, should the present sources be worked out."

I listened to Lang Hancock's son-in-law talk about how rich the Pilbara iron deposits were, for the seventy-five hours he and

I were in the air together. Official reports today say that Western Australia possesses 29 percent of the world's iron reserves. It's a banded iron formation. The iron precipitated out of salt water over two and a half billion years ago.

So when Quinton started talking to me about how he believed the source of the gold in the Witwatersrand was its precipitation in the presence of oxygen produced by single-cell bacterial colonies, I became a believer. If the iron precipitated out of salt water in the presence of oxygen, and the gold precipitated out of salt water in the presence of oxygen, you cannot have one without the other.

Fifty years from now, after tens of millions of ounces of gold have been produced from the Pilbara, it will be as obvious to everyone as it was to me in 2008.

You see, it's not rocket science.

Novo has made their first gold pour at Nullagine as I release this book for sale.

To be continued...

REFERENCES

1. https://en.wikipedia.org/wiki/Bre-X

2. https://en.wikipedia.org/wiki/4-H

3. https://en.wikipedia.org/wiki/Roebourne%2C_Western_Australia

4. https://fremantlestuff.info/people/angelo.html

5. https://en.wikipedia.org/wiki/Emma_Withnell

6. https://en.wikipedia.org/wiki/Lang_Hancock

7. https://www.thefreedictionary.com/pommy

8. http://y20australia.com/rab-drilling-mining-explained/

9. https://en.wikipedia.org/wiki/De_re_metallica

10. https://en.wikipedia.org/wiki/Caribou,_Colorado

11. https://paperity.org/p/4026623/witwatersrand-gold-deposits

12. https://www.mountvernon.org/library/digitalhistory/digital-encyclopedia/article/first-in-war-first-in-peace-and-first-in-the-hearts-of-his-countrymen/

13. https://www.history.com/news/the-day-skylab-crashed-to-earth-facts-about-the-first-u-s-space-stations-re-entry

14. (directed to FT subscription page)

15. https://www.proactiveinvestors.com.au/companies/news/201

791/legend-mining-continues-to-draw-comparisons-with-nova-bollinger-deposit-201791.html

16. https://en.wikipedia.org/wiki/Dead_centre_(engineering)

17. http://www.321gold.com/editorials/moriarty/moriarty081512.html

18. https://www.novoresources.com/_resources/news/2012-08-13.pdf

19. http://www.321gold.com/editorials/moriarty/moriarty082712.html

20. https://en.wikipedia.org/wiki/National_Instrument_43-101

21. https://www.voiceamerica.com/Show/1501

22. https://www.northernminer.com/news/newmont-takes-35-7-stake-in-novo-resources/1002578301/

23. https://en.wikipedia.org/wiki/Bulk_leach_extractable_gold

24. https://www.novoresources.com/_resources/news/2013-12-17.pdf

25. https://www.novoresources.com/_resources/news/2014-03-04.pdf

26. https://www.novoresources.com/_resources/news/2014-03-19.pdf

27. https://www.novoresources.com/_resources/news/2014-07-24.pdf

28. https://www.novoresources.com/_resources/news/2014-08-28.pdf

29. https://www.novoresources.com/_resources/news/2014-12-10.pdf

30. https://www.businesswire.com/news/home/20140512006751/en/Newmont-Signs-Agreement-to-Sell-Jundee-Underground-Gold-Mine-in-Australia

31. https://www.amazon.com/Games-People-Play-Psychology-Relationships/dp/B0007DYNTE/ref=tmm_hrd_swatch_0?_encoding=UTF8&qid=1602437696&sr=1-1

32. https://en.wikipedia.org/wiki/Volcanogenic_massive_sulfide_ore_deposit

33. https://www.geographie.uni-wuerzburg.de/fileadmin/04140600/WR_BKGR/Frimmel_Wits_SEG_SP18_2014.pdf

34. https://www.springer.com/journal/126

35. https://www.researchgate.net/publication/273278727_First_whiffs_of_atmospheric_oxygen_triggered_onset_of_crustal_gold_cycle

36. (directed to Northern Star subscription page)

37. https://www.hecla-mining.com/midas/

38. https://www.novoresources.com/_resources/news/2014-12-10.pdf

39. https://www.novoresources.com/_resources/news/2015-02-09.pdf

40. https://www.novoresources.com/_resources/news/2015-03-26.pdf

41. https://www.novoresources.com/_resources/news/2015-04-02.pdf

42. https://www.novoresources.com/_resources/news/2015-05-26.pdf

43. https://www.theglobeandmail.com/report-on-business/investor-eric-sprott-stepping-down-as-chairman-of-sprott-inc/article34630762/

44. https://www.globenewswire.com/news-release/2015/05/11/1279060/0/en/Newmarket-Gold-and-Crocodile-Gold-Merge-to-Establish-a-New-Platform-for-Gold-Asset-Consolidation.html

45. https://finance.yahoo.com/news/crocodile-gold-reports-drilling-results-220000972.html

46. https://www.spglobal.com/marketintelligence/en/news-insights/trending/3s62jiinj9202_qd-e0a0w2

47. https://www.kl.gold/news-and-media/news-releases-archive/default.aspx

48. https://www.novoresources.com/_resources/news/2015-08-17.pdf

49. https://www.novoresources.com/projects/pilbara/blue-spec/

50. https://www.novoresources.com/_resources/news/2016-08-25.pdf

51. https://en.wikipedia.org/wiki/The_Treasure_of_the_Sierra_Madre

52. https://www.dmp.wa.gov.au/Documents/Minerals/132298_Mining_Notice_Special_Pros_Licence.pdf

53. https://en.wikipedia.org/wiki/Tiny_Rowland

54. https://www.perthnow.com.au/business/gold/meet-mick-shemesian-the-mystery-man-behind-the-pilbara-gold-scramble-ng-b88662343z

55. https://hotcopper.com.au/threads/ann-new-style-of-gold-mineralisation-for-west-pilbara-at-purdys.3237395/

56. https://www.novoresources.com/news-media/news/display/index.php?content_id=232

57. https://www.youtube.com/watch?v=SiAGlWyjRq8

58. https://en.wikipedia.org/wiki/Peanut_gallery

59. https://www.novoresources.com/news-media/news/display/index.php?content_id=232

60. https://hotcopper.com.au/search/34225434/?q=%2A&t=post&o=relevance&c%5Bvisible%5D=true&c%5Buser%5D%5B0%5D=194432

61. https://hotcopper.com.au/threads/why-no-announcement-from-arv.3598161/page-66

62. https://hotcopper.com.au/threads/possession-is-9-10-ths-of-the-law.3610566/

63. https://hotcopper.com.au/threads/my-estimate-of-novo-sp-

fair-value.3610662/

64. https://hotcopper.com.au/threads/possession-is-9-10-ths-of-the-law.3610566/page-9

65. https://hotcopper.com.au/threads/my-estimate-of-novo-sp-fair-value.3610662/page-8

66. http://www.321gold.com/editorials/moriarty/moriarty082317.html

67. https://www.novoresources.com/news-media/news/display/index.php?content_id=319

68. https://www.novoresources.com/news-media/news/display/index.php?content_id=324

69. https://www.investopedia.com/terms/s/strategic-default.asp

70. https://www.novoresources.com/news-media/news/display/index.php?content_id=339

71. https://www.novoresources.com/news-media/news/display/index.php?content_id=346

72. https://www.novoresources.com/news-media/news/display/index.php?content_id=352

73. https://www.novoresources.com/news-media/news/display/index.php?content_id=355

74. https://www.novoresources.com/news-media/news/display/index.php?content_id=361

75. https://www.novoresources.com/news-

media/news/display/index.php?content_id=363

76. https://www.novoresources.com/news-media/news/display/index.php?content_id=366

77. https://www.australianmining.com.au/news/millennium-minerals-moves-into-voluntary-administration/

78. https://www.novoresources.com/news-media/news/display/index.php?content_id=376

79. https://www.novoresources.com/news-media/news/display/index.php?content_id=379

80. https://www.novoresources.com/news-media/news/display/index.php?content_id=380

81. https://www.novoresources.com/news-media/news/display/index.php?content_id=381

82. https://www.novoresources.com/news-media/news/display/index.php?content_id=384

83. https://www.novoresources.com/news-media/news/display/index.php?content_id=387

84. https://www.novoresources.com/news-media/news/display/index.php?content_id=394

85. https://www.novoresources.com/news-media/news/display/index.php?content_id=399

86. https://www.novoresources.com/news-media/news/display/index.php?content_id=402

87. https://www.novoresources.com/news-media/news/display/index.php?content_id=419

88. https://en.wikipedia.org/wiki/Lang_Hancock

89. https://en.wikipedia.org/wiki/Harry_Page_Woodward